Thinking About
Biology

Stephen Webster
Science Communication Group
Imperial College London

CAMBRIDGE
UNIVERSITY PRESS

PUBLISHED BY THE PRESS SYNDICATE OF THE UNIVERSITY OF CAMBRIDGE
The Pitt Building, Trumpington Street, Cambridge, United Kingdom

CAMBRIDGE UNIVERSITY PRESS
The Edinburgh Building, Cambridge CB2 2RU, UK
40 West 20th Street, New York, NY 10011-4211, USA
477 Williamstown Road, Port Melbourne, VIC 3207, Australia
Ruiz de Alarcón 13, 28014 Madrid, Spain
Dock House, The Waterfront, Cape Town 8001, South Africa

http://www.cambridge.org

First published 2003

Printed in the United Kingdom at the University Press, Cambridge

Typefaces Lexicon No. 2 9/13 pt. and Lexicon No. 1   *System* LaTeX 2$_\varepsilon$ [TB]

A catalogue record for this book is available from the British Library

Library of Congress Cataloguing in Publication data

Webster, Stephen, 1957–
    Thinking about biology / by Stephen Webster.
        p.   cm.
    Includes bibliographical references (p.).
    ISBN 0 521 59059 0 – ISBN 0 521 59954 7 (pb.)
    1. Biology.   I. Title.
    QH307.2 W44 2003
    570–dc21   2002073599

ISBN 0 521 59059 0 hardback
ISBN 0 521 59954 7 paperback

For Giovanna, Lorenzo and Luca

Contents

Acknowledgments

My first debt is to Paddy Patterson, of the University of Sydney, whose commitment to biology extends fully to his teaching. At different times the students of Quintin Kynaston School, King Alfred School, Birkbeck College and Imperial College London have reminded me why the philosophical questions of science are important. Many colleagues have encouraged me, rooted out errors or sharpened my prose: John Barker, Dawn Moore, John Peisley, Stephen de Brett, Terry Preston, Nick Russell, Joan Leach, Peter Forey, Richard Fortey, John Dupré, Ken Arnold, Jeff Thomas, Denis Wright, Michael Reiss and an anonymous referee for Cambridge University Press. Especial thanks are also due to the many restaurants and cafes of South Kensington, hosts to innumerable and enjoyable editorial sessions with Ward Cooper, my editor at Cambridge University Press. As always my greatest debt is to Giovanna Iannaco, who provides zeal and criticism at all the right moments.

Introduction

At some point in their work, most biology students will ask themselves: why is there so much to learn? Though the deeper principles of science may be valuable, they are always in danger of being overwhelmed by the mass of detail that fills all textbooks, and attends all courses. For students and teachers alike, undergraduate courses in the biological or life sciences can easily become exercises in fact management. This is a problem for the many students interested in exploring how their subject relates to wider contexts, or who are uneasy at the narrowing choices they had to make in order to study any of the biological sciences at university. Others, who want their biology degree to be as useful as possible, need to see that science can be discussed in a way that non-scientists find compelling and important. For those who choose a career in research, the necessarily strict disciplines of laboratory life are likely to need complementing with a feeling for the philosophy of the subject: a knowledge of how biology gains its authority, how it presents itself in public, and how it relates to other sciences, to the arts and to the humanities.

The purpose of this book is to make these connections. It aims to help students find some new intellectual perspectives on their studies. This is why the book is called *Thinking About Biology*. Throughout, I try to illuminate two kinds of connection. The first set explores the links between different parts of biology, emphasising the relations between evolution and genetics, between cell theory and the techniques of microscopy, and between organismal biology and molecular biology. These connections have a simple purpose: to remind the student that the different modules they may compile during a biology course do indeed link up – they are part of the same discipline. My second set of links are more unusual for a biology textbook, because they extend far beyond the conventional bounds of a life sciences course. Here I explore our subject's philosophical foundations, its relationship with politics and ethical discussion, and its representation in the media. These are topics that are interesting to most students and teachers, and are important tools for anyone wishing to put their work to use. Their obvious value as an accompaniment to, and motivation for, the study of biology is being increasingly recognised by the more enlightened schools, colleges and universities.

Thinking About Biology is a textbook because I ground all my discussions in material likely to be found in any life science or medical curriculum. I start with the biology, and then look further afield. Thus the book is not an introduction to the history and philosophy of science, though it goes some way in that direction. Amongst my diverse set of topics and arguments, a few broadly philosophical themes recur. For example, I explore the way a topic as fundamental as the cell theory has an argumentative and fraught history, one that suggests facts are not timeless discoveries, but are rather more fragile and dynamic than that. Similarly, though one can make a conscious decision to view modern molecular genetics simply as a set of technical achievements, *Thinking About Biology* explores the way such achievements relate to the public debate about the limits of science. An important theme of the book is that the well-trained graduate knows how science impacts on society, understands that science itself is affected by society, and must be responsive to wider debates.

In short, this book is a practical manual for the thinking student. It provides just a few of the tools needed to become a reflective, as well as a technically proficient, practitioner of the craft of biology. As the book puts such an emphasis on the concept and desirability of the 'reflective scientist', I should briefly flesh out why I consider this important.

No one could oppose the idea that the learning of biology should be a thought-provoking exercise. Obviously, biology classrooms, lecture halls and laboratories are filled with people trying to make sense of nature. The problem is that the academic environment may hinder, rather than enhance, the intellectual spirit. For the students, courses may be too fragmented, too laden with factual content, or too heavily assessed. Teachers face pressures too. College and university lecturers are themselves assessed, tests based in many cases on where, and how often, their research is published. The institutions where they work are as likely to compete as to collaborate, and find themselves anxiously occupied with their ratings in league tables. All of this makes it much more difficult to provide in formal education the open-ended and meandering discussions needed for encouraging reflection.

Yet the need for such discussions is very great. Firstly, it is absolutely clear that students enjoy the chance to discuss and debate their subject, and want more of these opportunities. Secondly, a glance beyond the university and college corridors quickly reveals that a great deal of science is being discussed in the press, on the broadcast media and through the Internet. Some scientists consider this debate largely misinformed, usually sensational and inevitably oversimplified. Others realise that so great a public interest in such matters as genetic modification, bovine spongiform encephalopathy (BSE) and xenotransplantation, is largely a good thing – and that it would be mad for scientists and students simply to ignore it. However, to take part in the debate and to make a contribution, we will need to avoid all arrogance and be prepared to listen. Nor need this simply be a buttoning of the lip. The philosophical and historical themes explored by *Thinking About Biology* show science to be a fallible and human endeavour, one incapable of establishing final truths. Instead, just like the rest of life, it is argumentative, wandering and personal.

The book has two parts, Chapters 1–4 and Chapters 5–8. The first half is longer and embarks on a study of some relatively conventional areas of the philosophy of biology. Chapter 1, **Facts?** uses the history of cell biology as a vehicle for exploring biological methods and their reliability. Chapter 2, **Reductionism**, follows aspects of the well-known anxiety surrounding biology's tendency to understand organisms by first breaking them up. Chapter 3, **Evolution**, is an area that is traditionally the home for literary and thoughtful biologists. Chapter 4, **Biology and animals**, ventures into animal rights, an area that seems appropriate for students to discuss, but unfortunately remains a somewhat tight-lipped debate in biological circles. The second half of the book takes on wider contexts. Commentary on scientific controversies, for example the public rows about BSE (mad cow disease), genetic modification and xenotransplantation, forms Chapter 5, **Controversies in biology**. The ethical implications of the Human Genome Project are discussed in Chapter 6, **Making sense of genes**. Chapter 7, **Biology and politics**, looks at some links between politics and biology, such as the history of eugenics. The book finishes with Chapter 8, **Research ethics**, in a consideration of this contemporary topic that takes as its centre a debate about the honesty of scientists.

How should you read the book? Most importantly, remember that *Thinking About Biology* aims to be thought provoking rather than authoritative. There is no need to read the chapters in order. Like any textbook, you should use the Contents section and the Index to find areas that interest you, or are related to your studies. Each chapter is made up of a few sections, each designed to be read reasonably easily, and I hope enjoyably. You will find comments, references and suggestions for further reading in footnotes throughout the book. References introducing the philosophy of biology are given in the short list below.

Further reading:

Chalmers, A.F. (1999). *What is This Thing Called Science?* 3rd edn. Buckingham: Open University Press.
Ruse, M. (ed.) (1989). *The Philosophy of Biology*. New York: Macmillan.
Sterelny, K. and Griffiths, P.E. (eds.) (1999). *Sex and Death: an Introduction to Philosophy of Biology*. Chicago: Chicago University Press.

Facts?

1.1 The problem with cannabis

In Amsterdam, they say, you can approach a policeman and ask the best place for buying cannabis. Very likely you will be courteously pointed to one of the city's 'coffee shops', where marijuana in a number of forms is on sale, to be enjoyed along with coffee and newspapers. The legalisation of cannabis in the Netherlands is 'The Dutch Experiment', and is a focus of interest for the interminable arguments about drug control in other countries. The liberal Dutch attitude contrasts with the stricter attitudes of the authorities in the UK, where until recently cannabis use was an arrestable offence, with 300 000 people street-searched each year and 80 000 arrested. The differing attitudes of the European countries to drug use is one reason for the constant newsworthiness of cannabis. Another reason is its widespread use. Some 50% of British 16–19 year olds have smoked cannabis; across Europe, there are 45 million regular users. The controversy takes various forms. Some argue that cannabis should be decriminalised. With this strategy, possession remains an offence, but leads to a fine or a warning, rather than to prosecution and a criminal record. Others go further and call for legalisation,so that cannabis is freely available, taxed and even supplied by the state. According to its advocates, legalisation of heroin and ecstasy, as well as of cannabis and amphetamines, will reduce the demand for drug dealers, and so reduce drug-related crime. Moreover, so the argument runs, when criminal suppliers are put out of business, the health problems associated with contaminated drugs will disappear too: government-controlled supplies will be quality assured.

In July 2002, the UK Labour Government confirmed that it was to re-classify cannabis, changing it from a Class B to a Class C drug, so that it will be in the company of mild amphetamines, tranquillisers and anabolic steroids rather than barbiturates, codeine and speed.[1] The change has a pragmatic element and was driven by a consideration of police priorities. Telling someone to stub out a joint takes 10 seconds; arresting and charging them takes 3 hours. No doubt establishment opinion is warming to the idea that cannabis is no more dangerous than alcohol or nicotine: politicians want to visit their undergraduate sons and daughters at college, not in jail. Changing views in the medical profession are also forcing a re-appraisal. For example, in 1998 a committee of the House of Lords (the UK parliamentary upper house) recommended that doctors should be able to prescribe herbal cannabis to people with certain illnesses, such as multiple sclerosis. According to the committee's report, the possible benefits patients might get from cannabis meant that it was wrong to expose such patients to legal action simply because they decided themselves to use the drug to alleviate symptoms. Scientists too are involved in the debate over society's proper attitude to drugs. It might be, for example, that scientific research will establish more precisely when and how cannabis, or heroin, is dangerous. A government, facing calls for a change in the law, will ask the following questions: does cannabis use carry the risk of long-term personality change, does it reduce your aptitude to work, is it addictive? The experts called in to rule on the issue will be physiologists as well as the police, psychologists as well as head teachers.[2]

There have been many scientific trials trying to measure the short-term and long-term neurological effects of cannabis or its active ingredient, tetrahydrocannabinol (THC). In addition, scientists and psychologists have investigated whether and how cannabis is addictive. Finally, it should be possible for social scientists to confirm or refute rumours that cannabis is a gateway drug, steadily drawing its users towards a life of needles, addiction and social dysfunction. On the one hand, no one disputes

[1] Class A drugs include heroin, opium, crack, LSD and ecstasy.
[2] The British parliamentary Conservative party generated amusement during its 1999 conference when Anne Widdecombe, then the party's home affairs spokesperson, declared that once in power she would inaugurate a 'zero tolerance' policy towards cannabis. Under the new law, anyone found in possession of even the tiniest amount would automatically face a fine of £100. The policy was quickly dropped after five senior Conservative politicians revealed that they had smoked cannabis when they were students. Soon cannabis was to be reclassified, and it became common to hear police chiefs speculating about the positive effects of full legalisation. The magazine New Scientist has a useful archive on both the scientific and the political debates (http://newscientist.com/hottopics/marijuana/). See also the archive of articles on the topic maintained by the UK newspaper The Guardian (http://www.guardian.co.uk/).

the importance of the issue: if cannabis is dangerous, then people should be protected. On the other hand, if it is not harmful, or can even alleviate medical conditions, then people should not be jailed for growing it in the greenhouse. Yet, in spite of the science brought to bear on the issue, no final judgement on the safety of cannabis has yet emerged. Even the facts generated by the scientific research are disputed: there are plenty of research data, but no one can agree on what they mean.

Take the question of addiction. An American study at Baltimore's National Institute on Drug Abuse described caged squirrel monkeys becoming addicted to THC. The monkeys were given an injection of THC every time they touched a lever. Soon enough they were hitting the lever deliberately and giving themselves injections as often as 60 times an hour; the conclusion drawn is that cannabis is physically addictive. Meanwhile, the statistics from the Netherlands, where cannabis is decriminalised, are sometimes used to point to an opposite conclusion: that cannabis is not addictive. The percentage of Dutch people who use cannabis is lower than in many other European countries, including Britain. Moreover, the number of Dutch drug addicts has not increased; in fact their average age is rising, showing that young cannabis smokers in the Netherlands are not moving onto something harder. The problem for campaigners on both sides is that the statistics do not close the argument. Neither the data from the Netherlands (done by survey of people's behaviour), nor the data from the Baltimore experiment (done by laboratory work on monkeys), are conclusive. Instead of producing useful predictions for people's behaviour and physiology in a wide variety of situations, the Dutch and Baltimore studies may simply tell us something about people in Amsterdam, and monkeys in Baltimore.

Apart from the question of addictiveness, one of the particular concerns about cannabis is that it lowers mental performance. Once again, science finds it hard to rule one way or the other. There are claims that cannabis users do worse at school and college, and are more likely to become delinquent, but the evidence for this is disputed. For example, there are trials where heavy cannabis users are asked to refrain from smoking for some days, and then to undergo manual and intellectual tests. In a study at Harvard Medical School, individuals who had smoked more than 5000 joints agreed to abstain and then take part in some computer games. They were found to be more aggressive than a group of light smokers. This, however, does not prove long-term damage, but perhaps only the irritation caused by withdrawal symptoms. Moreover, people who become aggressive in laboratory trials will not necessarily be violent in the real world. Even if it was shown that cannabis users underperform in class, this would not necessarily pin down the drug as to blame. Perhaps people who fail at school are also more likely to use cannabis. The old stereotyping of cannabis users as lazy, or underachieving at college, or unable to maintain relationships, are not likely to be judged true or false by simple scientific trials. The problem is distinguishing between cannabis as a cause, and cannabis as an irrelevance. One in ten road accidents involve drivers with cannabis in their bloodstream, but many of these drivers have alcohol in it as well, and the way individuals vary in their response to cannabis simply is not understood. As a result of these kinds of problems, neither the effects of cannabis, nor its dangers, are reducible to a neat series of undeniable statements. The scientific research is not producing general truths.

The fact that the science does not offer certainty allows another factor to make a strong impact. This is the world of social and political opinion. Many people are horrified by the idea of cannabis being decriminalised. For them, it is simply a fact that cannabis is dangerous, causes college dropout, and inevitably converts our finest youth into comatose junkies. They would much rather someone drinks half a bottle of whisky, than smokes a joint. The fact that others consider alcohol more dangerous, more addictive and more socially ruinous, is an irritation mostly ignored. Clearly, prejudice is at work here. Could prejudice affect the interpretation of scientific results, turning the data in a particular direction, or in none? Cannabis researchers may be looking for particular results. The availability of money may determine whether research is done in the first place, and who does it. Opinions affect whether research is carried out, how it is received, and even whether it is published. The conclusions of the House of Lords report, though based on sifting through the scientific evidence available, were sidelined by the UK Government, who announced that they would wait for more conclusive evidence to emerge. More dramatically, when the World Health Organization compiled a report comparing the dangers of cannabis with those of alcohol and tobacco, and this showed that cannabis is the least dangerous of the three, political pressure led to the report remaining unpublished.

Summary: the facts of cannabis

Cannabis contains a chemical that affects the body. Many claims are made about the dangers of cannabis – to individuals and to society. With so many people buying and smoking cannabis in defiance of a hostile establishment, it is important to research the truth of these claims. The scientific tools for this research include neurophysiology, psychology and sociology, but we have seen that science is not able to close the argument: its data are disputed, and its interpretations vary. It is a common assumption that the particular merit of science is that it is one area of life where proof and certainty are guaranteed. The cannabis debate suggests something else: that science does not provide final answers and definitive proofs, but rather, that all science involves dispute, and that all science is fought over. This is true not only of the science of cannabis, but of every area of biology too.

1.2 The making of the cell theory

I started this chapter by discussing cannabis. I emphasised how hard it is to find clear evidence on the safety of cannabis. Clearly, social prejudice is a powerful force in determining the history of legal attitudes to cannabis. I discussed too the way that scientific research also finds it hard to avoid dispute and equivocation, and I suggested that this ambiguity, or at least lack of certainty, is a core feature of all of science, not only of admittedly complex physiological interactions. In this section, I take the argument further by looking at cell biology, a much more traditional and mainstream area of biological research than tests on cannabis addiction in monkeys. Cell theory, like evolutionary theory, is a well-established field that forms the basis of all biology courses, and of all biology. Surely this is a field so well understood that it has long since settled into a middle-aged complacency, with everything determined except for a few minor upsets here and there. I will suggest instead that here too, uncertainty and dispute are a central theme. My aim is to raise in your mind the idea that biology is more dynamic, and less fact oriented, than some of your textbooks, and your teachers, may suggest. In particular, I will look at the history of research into what we now call fertilisation – the fusion of sperm and egg – and try to show how a basic biological idea was itself the product of much confusion and disagreement. However, I do not want to imply that all the disputes took place a long time ago, and by using some examples from contemporary cell biology, I hope that you will see that uncertainty and lack of knowledge are fundamental aspects of the modern scientist's life.

Behind the daunting detail of a cell biology textbook lies something simple and fundamental. I refer, of course, to the cell theory itself: the profound concept that all living things are composed of cells, that all cells come from earlier, pre-existing cells, and that all organic material in nature has been formed by cells. Yet this basic rule of biology was not established merely as a result of the invention of microscopy and the first observations of tissue fine structure. There is a gap of 174 years between the first description of box-like units in cork (1665), and the confident assertion of the cell theory (1839). The pioneer microscopist was Robert Hooke, who examined slices of cork, and was reminded of cells – the places where monks sleep and pray; but he did not immediately suggest that all tissue is made of cells, or comes from cells – why not? The answer is that cell theory had to be made, a net of ideas had to form. It was not simply a matter of looking down a microscope at plant material, finding square structures, and instantly realising that cells make up all tissue, divide, and have different parts. It was not just a blinding flash of inspiration. A great amount of thinking and arguing, as well as looking down microscopes, would be needed before cells, at least as we conceive them, could be seen. Microscopes were needed to make the structure visible; but to make sense of that structure, you need to think, and to have arguments. Those arguments in turn influence how the microscope is used, and what is observed. It is this mix of looking and thinking that makes doing biology a creative process, not simply a cataloguing of facts. It is in this sense that cell theory was created, not discovered.

In order to develop further the creativity of making science, I will now concentrate on one type of cell, and its intellectual history: the reproductive cell – gametes, or sperm and eggs. As with cell theory in general, there was a huge gap in time between the first observation of sperm under the microscope, and their conceptualisation as partners in fertilisation. Sperm were first observed under the microscope in 1670. Yet the idea of fertilisation as a process that puts together inherited material from two parents, dates only from 1870 – a 200-year interval. This delay in reaching the modern understanding was not simply a matter of waiting for better microscopes: a great deal of thinking had to happen too.[3] Some of that thinking we now find strange: one nineteenth-century biologist, von Baer, thought that spermatozoa were parasitic worms swimming in the semen.

---

[3] Historians of science strongly dislike accounts of science that see the work of previous centuries as slowly clearing mists of ignorance. It is easy to characterise past scientific knowledge as simply a catalogue of mistakes. Historians point out that it is too simplistic to use the 'spectacles' of our modern understanding as a technique for judging the work of earlier scientists. This discredited historiographical method is known as 'Whiggish history'. Such accounts of the past are distorted by being filled out with recognisable ancestors to our intellectual world. Ideas that we now make no use of are simply stripped out, or condemned as absurd. As a result, the history becomes an unreliable account of the debates and intellectual battles that were actually taking place.

However surprising this idea seems to us now, it does remind us that the observation of small motile objects in semen does not in itself amount to a discovery that these wriggling things were needed for reproduction. It was obvious that sexual intercourse is needed to make babies, and that ejaculated semen is the vital male component; but what exactly was this thing that the male supplied – nutrition, or heat, or a mysterious force, perhaps electrical? The basic function of the male semen had been perfectly captured in the Old Testament story of Onan in the book of Genesis. There, Judah orders Onan to have sexual intercourse with (and make pregnant) his brother's widow Tamar (Judah's daughter), in accordance with levirate law. Reluctant to help out his brother in this way, Onan attempted subterfuge by practising coitus interruptus, or to put it in the words of Genesis: 'Then Judah said to Onan, "Go in to your brother's wife, and perform the duty of a brother-in-law to her, and raise up offspring for your brother." But Onan knew the issue would not be his; so when he went in to his brother's wife he spilled the seed on the ground, lest he should give offspring to his brother. And what he did was displeasing in the eye of the Lord, and he slew him.' Onan's sad end is marked by the word 'onanism', a simile for masturbation – but in this case it was Onan's disobedience of God's law, not the act of masturbation, that proved the fatal mistake.

Human semen is a much more obvious thing than human eggs, and so there was early speculation on what and where might be the corresponding 'seed-stuff' within the woman. The ancient Greek philosopher Aristotle was interested in the question and declared that the role of semen was to act upon the menstrual blood, fashioning it into a baby. These and other ideas floated up through the centuries; a more modern scrutiny followed the development of the microscope and the discovery by Antony von Leeuwenhoek of 'spermatic animalcules'.

Antony van Leeuwenhoek is the most famous, though not the first, of the early microscopists. His interest in microscopy was provoked when he saw the illustrations of a completely new microscopical world, as revealed in a revelatory book, *Micrographia* (1665). It was in this book, alongside drawings of magnified full stops and pin heads, that the English physicist Robert Hooke had described the monkish compartments – cells – that he saw inside cork. For Hooke, the cell was empty and inactive: far from the boiling turbulence that is evoked for modern biologists by the word 'cell'. Leeuwenhoek dramatically improved the magnifications available by simplifying the optics, choosing to build microscopes with only one lens instead of two. Leeuwenhoek was so miraculously expert that with one expertly ground lens, and his own presumably superb eyesight, he could achieve working magnifications of ×200. The cells he saw were more dynamic than Hooke's, more recognisably alive. He was a fine letter writer, and the reports he sent over to the Royal Society of London (founded in 1660) are a reminder of how vivid technical writing can be. Here, for example, is a description of the green alga *Spirogyra*, found in a local lake: 'Passing just lately over this lake ... and examining this water next day, I found floating therein divers earthy particles, and some green streaks, spirally wound serpent-wise, and orderly arranged after the manner of the copper or tin worms, which distillers use to cool their liquors as they distil over'. Thus, in the liveliest of prose, Leeuwenhoek introduced to the reading public such things as blood cells, microscopic nematodes, and, during the 1670s, the 'spermatic animalcules'. These were what we now call spermatozoa.

Thinking about sperm and eggs

The function of sperm was as unclear as their interior. Though the sperm had been described, no one as yet had seen a mammalian egg. However, supplementing the few (and varying) microscopical descriptions there came a rich mix of expectations, both scientific and social. Leeuwenhoek was vigorously opposed to the idea of spontaneous generation – the belief that the decay of plants and animals produced new life in the form of worms and insects. He mobilised all his microscopical discoveries to show that life could be very small indeed, even though invisible to the eye, and argued that the tiny intricate components he saw inside worms and insects could come only from life, not from putrefaction. Leeuwenhoeks's discovery of spermatozoa was important to the campaign. Though the mechanism was unknown, a reproductive role was suggested by the fact that, with care, spermatozoa could be found in the semen of any mammal. Leeuwenhoek favoured 'preformationism', the belief that the embryonic animal contains, in miniaturised form, all the adult organs, which gradually enlarge and become visible as the embryo develops. The concept was applied also to eggs (ova), and to sperm: they too could be a tiny storehouse of preformed parts. However, there was disagreement on whether the miniaturised organism would be in the sperm or in the ovum. Social factors may have contributed to a temporary dominance of the sperm as the home of the embryo. It is men who make sperm, and generally, it is men and not women who inherit titles and fortunes. 'Animalculist preformationsim', the embryo-in-the-sperm, was the biological manifestation of one of society's most rigid prejudices, that power and influence pass down the male line. With such a view, the female contribution to making babies can only be in providing a nourishing home for the little baby, whether folded up inside the sperm, the womb, or the cot. The sperm, in other words, contained a preformed person, which would be able to grow as soon as implanted in a woman. The female role in such an account does not include inheritance, but does include nurturing. This is an example of science echoing society. All this is dramatically illustrated by Hartsoeker's 1694 drawing of a perfectly formed man or 'homunculus', arms and legs folded, miniaturised but recognisably human, sitting inside the spermatic animalcules.

However, there was opposition to Hartsoeker's homunculus. If every sperm carries a little man, then there must be millions of them, but even the most active father would be hard-pressed to manage more than a dozen offspring. This certainly represented a waste of valuable male heritage. Meanwhile, by the beginning of the eighteenth century, dissection established that mammals develop from an egg. Preformationists therefore began to favour the idea that it was the egg which must contain the perfect, preformed person. This idea, ovism, became the dominant model for preformationists. The theory still fitted well with contemporary ideas of the universe. The seventeenth century physicist Isaac Newton, famous today for his three laws of mechanics, had described a clockwork universe, made by God but understood – and celebrated – by scientists. Newton demonstrated that the physics of terrestrial mechanics and celestial movements were the same. In this view, both God and science have a role. God makes and fits the minute cogs together and sets them running; science locates the cogs and describes their movement. The idea supported preformationism. Perhaps the egg, like the solar system, is a kind of machine, whose pre-squeezed components unfold and grow as the individual develops.

The freshwater polyp hydra was an important element in the eventual demise of preformationism. In 1741, Abraham Trembley watched hydra, described its cartwheeling walk and contractibility, and so showed it to be an animal, not a plant. Then, on further investigation, he found something really sensational: cut a hydra in two and two new animals regenerate. This did not fit well with ideas about preformation. If everything is preformed inside eggs, how can you have whole new organisms being created simply out of ordinary chopped-up animal tissue? Here was reproduction that involved neither egg nor sperm. It had been a central feature of ovism that eggs are the only tissues able to generate descendents. Trembley's descriptions were hard blows to the theory of preformationism, but did not destroy it completely, for as is well known, one can always ignore bad news. According to Trembley, the scientists of the Royal Society were themselves slow to see the significance. He remarked: 'The singular facts that are contained in the history of these small animals are the admiration of a great many people: but several people have been hesitant to admit them. There are those who have even said that they will not believe it when they see them. Apparently these men have some cherished system that they are afraid of upsetting.' Yet the 'cherished system' of preformationism was indeed opposed by another belief about embryology: epigenesis. This is the belief, emerging in the eighteenth century, that tissues and organs form gradually from an initially undifferentiated mass. During development primitive jelly-like material simply begins to acquire structure: an eye forms, a heart appears, a wing bud emerges. Under the microscope, as time goes by, the detail emerges. According to the epigeneticists, the details become visible slowly because they are forming from translucent living matter, initially devoid of structure. According to the preformationists, the detail becomes visible because the tiny invisible structures finally get large enough to be seen.

We do not need to go into the details of the argument between these two camps, but two points are worth noting, because they are relevant to many other scientific debates. Firstly, there is the question of evidence: what observations might definitely sway the argument in one direction, for example from epigenesis to preformationism? The two theories fitted the observations equally well. Those who believed in epigenesis argued that there was no evidence of preformed parts existing in the early embryo: nothing could be seen. Those who believed in preformation argued that there was no evidence that form was derived simply from jelly: though not visible, the preformed parts were surely there. Secondly, in this dispute, the lack of what we would call scientific evidence was amply compensated for by the robust intervention of belief and expectation. The preformationist saw the hand of God in the formation of all embryos, all at once at the Creation; once made, the mechanical unfolding of embryonic forms simply revealed God's divine purpose. The epigeneticists, in contrast, saw in the coagulation of jelly into tissues a greater triumph for the laws of nature: that such laws can create living matter, not just keep the cogs of the universe turning. Not that the epigeneticists were atheist: they believed in God, but did not see him as having to set everything off right at the beginning of the world. Epigenesis revealed how God was present in every Law of Nature, and had no need to make a special creation of every embryo: embryos formed naturally from undifferentiated matter.

I do not mean to suggest that developments in science are simply a result of wider changes in attitude. Indeed, just as the epigeneticists won the argument, a whole series of changes took place in the capabilities of scientists. One major change was in the microscope, now capable of finer resolutions and greater magnifications. However, a key factor in the adjustment of scientific minds to the plausibility of epigenesis was change in the wider intellectual context. Towards the end of the eighteenth century, the Romantic Movement in Europe reacted against the idea of a mechanical universe. In England, the great romantic poets included Samuel Taylor Coleridge and Percy Bysshe Shelley. However, in Germany, the movement was represented most strongly by philosophers such as Herder, Schelling and Hegel. Their descriptions of a progressive, living, universe clashed with the cold and automatic mechanics of Newton. In science the influence would be strongest where Newtonian mechanisms had made the smallest headway: biology. The influence was felt particularly in the debate between the preformationists and the epigeneticists. The great German philosopher Immanuel Kant, himself interested in science, stated in his *Critique of Judgement* (1790) 'Absolutely no human reason can hope to understand the production of even a blade of grass by mere mechanical causes'. Instead, according to Kant, one should simply take it for granted that life is self-organising and self-regulating. Epigenesis, descriptive of organic matter becoming more organised, fitted this philosophy well.

Political and social forces would also support epigenesis. The Enlightenment, a wave of ideas in Europe at around the end of the eighteenth century, saw hope in the power of human reason and doubted the value of religion and tradition. Human thought – an exemplar is science – would transform the world from backwardness and lawlessness to a place of justice and progress. Clearly, this resonates more completely with epigenesis than with preformationism. The former evokes images of jelly-like masses transforming themselves into ordered cells: progress. Preformationism, in contrast, suggests that we are in the grip of fate, with life unfolding itself remorselessly without any possibility of improvement or even real change. In this intellectual climate, scientists who supported epigenesis no longer had to defend themselves: the self-organising abilities of life were now taken for granted. The task now was to use better optics to improve descriptions of development, from the earliest stages. A new way of working had been inaugurated, with a new set of preconceptions, and if the early stages of life no longer consisted of a miniaturised adult, then the real role of sperm and egg could be given new scrutiny.

If you consider your own understanding of fertilisation, you very likely regard it a matter not simply of the fusion of two nuclei but, more crucially, as the bringing together of two sets of chromosomes. For a modern biologist, it is simply impossible to think of a sperm fusing with an egg without at the same time seeing a rearrangement of chromosomes, but in the eighteenth century, not only were there no ideas about chromosomes, there was also no stable scientific sense about cells or inheritance. You can easily understand, therefore, that when questions began to be asked about the function of sperm and eggs, the answers given would depend not solely on the improvement of microscopes, but also on the changing views of cells, cell division and inheritance.[4]

When chromosomes were first seen, it was their involvement in cell division that was so obvious, not their role in inheritance. This is not so surprising. What was starkly visible was simply that chromosomes go through a complicated series of movements during division; perhaps that was their job – some kind of mechanism for getting cells to split into two. Yet for many biologists, a theory of inheritance ought to be based on what was known about cells. Cell contents should give clues about inheritance. Moreover, common-sense observations of children, who look a bit like both their parents, suggested an equal investment from sperm and egg. The nineteenth-century cell biologist Nageli asked himself the simple question: with the sperm so tiny, and the egg so big, how could the two of them ensure an equal contribution of the 'inherited material'? Nageli suggested that there must be a fraction of the egg which is important in inheritance; he proposed calling this fraction the 'ideoplasm'. He thought that the ideoplasm could only be a fraction because the egg is 1000 times bigger than a sperm, and yet makes an equal contribution to the offspring. The part of the egg contributing to the offspring cannot

---

[4] I am emphasising here a fundamental aspect of the history of science – that scientific discovery is not simply a matter of improvements in technique and of new observations. There is always a context to scientific work, influences that extend far beyond what is usually called 'the world of science'. It is in this sense that scientific theories can be seen as created rather than discovered. The unpicking of the many factors that make up a scientific change is highly complex, and my description of seventeenth- and eighteenth-century biology is necessarily brief and selective. For a more in-depth account of the historical events outlined in this chapter, try Ernst Mayr's *The Growth of Biological Thought* (Cambridge, Mass.: Harvard University Press, 1982) and Shirley Roe's *Matter, Life and Generation* (Cambridge: Cambridge University Press, 1981).

therefore be any larger than the sperm. This excellent idea, a 'thought-experiment really', did not lead Nageli to make the apparently obvious connection: that the ideoplasm might be the nucleus.

The whole point about inheritance is that something passes down through the generations – but what? Perhaps at this time – around 1880 – ideas were beginning to link up. Microscopists could now watch fertilisation, and knew it to be some kind of interaction between the sperm and the egg. With Nageli's idea of the ideoplasm so influential, and the nucleus coming under closer scrutiny by an ever-growing community of cell biologists, some details began to emerge. The nucleus was patterned, not simply a blob of jelly. There were those large structures inside, which had proved so thought provoking: the chromosomes. Stains allowed their regular movement to be charted, and wondered at. Earlier in the nineteenth century, biologists liked to think of electrical excitation as the most important aspect of fertilisation. In this, they were simply being fashionable: the physicists of the time were revered for their laws of electromagnetism, and for their impressive, useful machines. Yet if fertilisation was simply a matter of a sperm electrically exciting an egg (as some physics-loving biologists had suggested) then why was cell division, and especially the movement of the chromosomes, so complicated? Alternatively, if the nucleus was merely a chemical storehouse, full of some kind of glutinous ideoplasm, why not simply divide it into two rather as you divide a bottle of wine? If equality of amount is the important thing, then you would expect a rather simple splitting, not this remarkable chromosomal dance. A theory was needed to put together all that was known about inheritance, cells and development.

Today we do not use the word ideoplasm. We say instead that the inherited material consists of particles called genes, arranged on the chromosomes, but in 1880, though Mendel had done his work and had indeed discussed inherited particles (he called them 'factors') his paper was languishing, unappreciated, and nothing was known of genes or DNA. So when a biologist called Roux, working at exactly the same time as Nageli, and thinking about the nature of the ideoplasm, worked out from first principles that the material must be particulate, the insight was crucial. Roux thought: if inheritance is by the passing on of particles, how do you make sure each of the daughter cells gets a full set? If each particle represents (today we would probably say 'codes for') a particular aspect of the living form, it is no good giving half the particles to one daughter cell, the other half to the second daughter cell; the result of this straight split would be to give each of the daughter cells a completely different set of particles, yet in mitosis the daughter cells are identical. Roux realised that for particulate inheritance to work each particle would have to double and the resulting pairs be split away from each other into two new cells. The best way to do this would be to have the particles lined up, like beads, and for the string to split longitudinally. Roux wrote: 'The essential process in nuclear division is the halving of each of the maternal corpuscles; all other processes serve the object to transfer one of the daughter corpuscles to the centre of one of the daughter cells, and the other to the centre of the other daughter cell.' Meanwhile, improving techniques in microscopy meant that the chromosomes could be studied not just during cell division, but also during fertilisation. A Belgian scientist, van Beneden, did the crucial experiment in 1883. Working on the nematode *Ascaris bivalens*, he showed that at fertilisation the two chromosomes of the male gamete join but do not fuse with the chromosomes of the female gamete, so that the zygote has four chromosomes.

So you can see that in the 1880s a whole series of ideas, new techniques and observations were coming together quite quickly to form an understanding that is recognisably modern. It must have been an exciting time. Theories about inheritance and about development had met with theories about cells. There was better microscopy and there were better conditions for scientific research. Each of these factors worked its influence, and a stable agreement amongst biologists became possible. The cycles of experiment and theory seemed to be leading to a coherent view of cells, chromosomes, and cell division and fertilisation: one that would last. As biology moved into the twentieth century, the foundations of cell biology had been laid: the nucleus is the carrier of inheritance; fertilisation is the fusion of two cells, but not the fusion of the chromosomes; there is a continuity of nuclei, from one generation to the next; there can be no break in this cellular continuity, and no possibility that a cell can form from anything other than another cell.

Summary: the making of cell biology

With few exceptions, cells are too small to be visible to the naked eye. In order to investigate cells, intervention with a microscope is required. The refinement of microscopes over 300 years has profoundly influenced biology, but it would be a gross simplification to imagine that the development of cell biology has simply been a matter of improving optics. A cell biologist works not only with cells and instruments, but also, as I have described above, with ideas.

Scientists, being human, speculate. They have imaginations, and see their work as creative. Therefore, however excellent the microscope, there will always be the need, and the desire, to interpret the shadows of a dimly seen image. The scientist's own preconceptions shape the interpretation. The scientist does not just read data passively; the information is sculpted and given meaning. It is this active searching – sculpting – that makes science a creative activity where social and intellectual contexts, personal whim and ambition, technique, skill and luck are each important. Together they make up 'the mangle of science'. To sum up, science is a complex social phenomenon, not just a technical activity. We expect to find in its discussions and its published papers plenty of facts and theories, but the production of these ideas is not simply a matter of accurate observations being carefully recorded. It is neither a list of discoveries made by modest workers, nor an expression of ambition, prejudice and academic manoeuvring; instead it is a mix of all these, and the mix is as creative in the twenty-first century as it was in the eighteenth century.

1.3 The edge of the amoeba

You might imagine that as the technique of microscopy improves, the possibility diminishes that scientists can argue about the meaning of observations, but you would be wrong. When electron microscopes were invented in the 1940s, magnifications of $\times 40\,000$ became possible. Ultrastructure, for example the detail of mitochondrial architecture, became visible. How can we be sure though, that those lines and dots and circles represent real biological structure, rather than lumps of precipitated stain or tears and folds in the specimen? Bear in mind the extremes of manipulation a biological specimen undergoes prior to being slid into an electron microscope's holding bay. Cells must be killed, dehydrated, fixed, stained, and sliced, before being positioned in a vacuum chamber and bombarded by a beam of electrons. Along the way, you can imagine, something might be added that fools the microscopist. Such changes, produced by a procedure rather than by nature, are called artefacts. Artefacts worry biologists, who want to take for granted that their complex techniques generate reliable data about cells. Thus, though it is true that biologists may be aware that they are active in giving shape to their data, they certainly do not want that active involvement to extend to the generation of false data, either deliberately or by accident. So biologists have to work hard to feel confident that the things they see do indeed exist.[5]

We can get a sense of the importance of artefacts in biology by looking at that great classic of cell science, the movement of amoeba. There was a time when every A level and High School biologist studied the internal motions of the amoeba's cytoplasm. These movements are so familiar that you might imagine that little remains to be found out about the way cytoplasm squeezes and pushes. The basic ideas were worked out at the start of the twentieth century. As a subject for investigation, cell movement has some obvious difficulties. How do you study the movement of cells if you cannot stain them, for fear of killing them? This was of particular relevance for scientists in 1970, who asked themselves the simple question: how does the amoeba grip? Strangely, when biologists worked to understand amoeboid movement, they only considered what was happening inside the cell. But whatever the internal wanderings of the cytoplasm, an amoeba is going nowhere if it cannot grip onto a surface. In other words, if an amoeba is to move, it must have traction. In the laboratory (though not in real life) amoebae move along cover slips or glass slides, suggesting that the friction – the traction – must be occurring in the zone between the amoeba membrane and the cover slip. This is an extremely thin zone; it is transparent, and any 'contact' between membrane and glass is going to be well below the theoretical resolution of light microscopy. Assuming that there is something interesting happening, how are you going to make it visible without killing the amoeba, and if you do find a technique for making it visible, how will you ensure against artefact?

The optical problems are huge. Although some amoebae are amongst the biggest cells in existence, they are transparent. Therefore, not much detail is seen using ordinary bright-field optics. The problem has long been solved by various breeds of interference microscopy, optical techniques that provide contrast by emphasising those tiny differences in density existing inside transparent cells, but not perceived by our eyes. Phase contrast microscopy is the most familiar of this stable of techniques, but there are many more, all exploiting the fact that if two wave fronts of light

---

[5] Philosophers have a name for most beliefs: 'realism' is the term used for the way most scientists believe their observations and theories describe real structures in the world, rather than (useful) figments of the imagination. Doubting realism is not the same as doubting the existence of the world, nor even that scientists discover truths. The philosophical debates over realism have been prompted by terms like 'electron' and 'quark'. These entities are not observable, yet they are important parts of physical theory, and are modelled successfully by mathematics. It is quite common in science and philosophy to find someone committed to the usefulness and even the truth of a scientific theory, but hesitant to declare the described entity (such as a quark) 'real'.

interfere after passing through a cell, destructive and constructive interference will provide a pattern of light and dark, a pattern that corresponds to real differences of density within the cell.

Having got the contrast, other tricks can be applied. Detail not picked up by the eye might be better served by highly sensitive cameras. Video can allow analysis of images at leisure, and such images can be analysed with computers. Every observation system creates 'noise' ranging from mess and scum smeared on lenses to the tough-to-eradicate problems like variations in background illumination, or small optical faults. A computer can store an image where there is only background illumination and no specimen, and compare it with a image where there is both background and specimen. Any detail in the specimen that also turns up in the specimen-less background image can be removed by the computer, cleaning up the image.

For a microscope to be useful it must not simply magnify, it must resolve, that is, see as distinct objects that truly are distinct. A good example can be taken from something studied earlier in this chapter: chromosomes. These structures are quite small, inside the nucleus, and the number depends on the species of organism. To see chromosomes it is no good simply magnifying them: you must be able to distinguish one chromosome from another. If all the chromosomes appear as one single blur, the simple magnification of that blur will not help at all. It is the resolution of tiny things – telling them apart – that is the aim of microscopists. There is a limit to a microscope's performance. When two objects cannot be told apart, the 'limit of resolution' has been reached. In light microscopes the limit of resolution is about 0.2 μm – half the wavelength of blue light. A typical cell might be 10 μm in diameter, and so can be easily resolved from its neighbour. A nucleus might be 1 μm across, and a vacuole 0.8 μm; with practice, and the right stain, these organelles can be seen with a light microscope at high magnification. Bacteria too, sized around 1 μm, are visible, but organelles such as mitochondria or the Golgi body stand at the edge of the invisible. The inner structure of these organelles, for example, the membrane convolutions so commonly seen in mitochondria, resist the light microscope completely. The description of this ultrastructural detail awaited the development of the electron microscope: the wavelength of an electron beam is far smaller than that of light and this is the most important reason for the exceptional resolving power of these remarkable machines – 0.1 nm.

For those wanting to discover how the amoeba got its grip, the key technology was another form of interference microscopy, reflected light interference microscopy (RIM). Complex in detail, the technique nonetheless relies on a simple principle. Light bounced off two or more partly reflective surfaces will show interference if the surfaces are very close together. Imagine an amoeba crawling along a cover slip. There are several surfaces: the surface of the amoeba and the two surfaces of the cover slip. The closeness of these surfaces, and the fact they are partly reflective, allows interference patterns to develop. To get the effect, light is shone up through the cover slip and into the amoeba. Most of the light shines straight through the glass and the amoeba and out the other side; this light is lost and plays no part in the experiment. Some light, however, shines through the glass but is reflected back by the underside of the amoeba. Other light is reflected straight away by the glass cover slip. It is these two wave fronts, one reflected from the amoeba, the other from the cover slip, which can be made to interfere and produce a pattern. The type of interference pattern depends on the distance between the membrane and the cover slip: a wide gap will give a certain level of brightness, a narrow gap another. In short, all the conditions for resolving tiny objects have been met and indeed RIM has the ability to distinguish between a gap of 20 nm and one of 100 nm, simply because different gap sizes produce different patterns. A big gap can be resolved from a small gap. The prediction of course is that the amoeba gets its traction by making close contact in some way with the substratum: by tiny feet perhaps, or by larger blocks of membrane? RIM will show where the contacts are, and give an idea of their shape.

In practice, to see the amoebae grip, they have to be made to crawl along in distilled water, which is difficult, as only some amoebae (for example *Naegleria*) can survive such conditions. If the water is saline, the image is grey and unpromising. As soon as the molarity of the saline solution is reduced enough, something extraordinary happens: black spots emerge vividly from the background, perhaps 15 to the cell. As the cell streams by, the spots stay in the same place: then, certainly, they must be attached to the substratum. The mathematics of the interference confirms the spots to be 'zones of extremely close contact' between the amoeba and glass. The cell uses these focal contacts as fixed points through which to gain purchase on the surface; the front of the organism puts down a few contacts onto the cover slip, and then flows past. Eventually, the focal contacts are at the back of the animal and are taken up into the cell. It is all very impressive, but how can we know that these observations of focal contacts

represent the real actions of amoeba, and are not merely artefacts resulting from extremely complex microscopy?

Even if focal contacts are shown to be real for laboratory amoebae
crawling over cover slips, biologists will want to be sure that they also exist for 'wild' amoebae living on a bumpy, uneven substrate.[6] In the traction experiments, cells moved through controlled aqueous solutions, and
when the salinity was reduced to zero, the focal points suddenly snapped
into focus; but amoebae do not inhabit distilled water – could the focal
contacts have formed simply as a result of the malign effect of the unfamiliar distilled water, or did they become visible only at that point, as a
result of the type of optics? If the former is true then the focal contacts are
simply an artefact of the experiment; if the latter is true then there is no
problem.

The best way of reassuring yourself about the problem of artefacts is to
use other sources of evidence. Suppose, for example, that you are observing cell structure using stains and the ordinary light microscope. Something apparently new appears; might it be an artefact? It will not be if it
can be seen using other types of microscopy, for example, low power transmission electron microscopy, nor if it appears, day after day, in different
preparations. Most people's first experience of science is at school. Here,
they are repeatedly told that experiments must be 'repeatable'.

In the case of the focal contacts of amoebae, independent lines of
evidence for their existence come from the fact that they have shown
up in transmission and scanning electron microscopy. A vital, further,
line of evidence involves biochemistry. Cytochalasin B is a drug with a
well-understood ability to interfere with the polymerisation of actin, the
molecule known to be the main component of the cytoskeleton. Focal contacts, if they are to act as anchor lines by which the amoeba pulls itself
along, must not only be firmly attached to the substratum; they must
also be integrated into the cytoskeleton. The hypothesis that focal contacts are part of normal life for amoebae – are how they get about – is supported by the following observation. Amoebae stop moving when they are

---

[6] This material is drawn from the work of Terry Preston (University College London), and from his
book *The Cytoskeleton and Cell Motility* (T.M. Preston, C.A. King and J.S. Hyams, Glasgow: Blackie,
1990). The field of cell motility is an excellent example of research that embraces both molecular
techniques and an interest in the life of the whole organism. The issue of how cells grip and move is
of obvious medical importance, for example, in understanding the spread of cancer, and molecular
techniques are necessary for the investigation. Preston's question 'how does this organism get around
in its world', is an important influence in determining the design of the chemical and microscopical
investigations. There is more on the virtues of thinking about the lives of organisms in the next
chapter.

in the presence of cytochalasin B. Internally the cytoplasm keeps flowing, but is unable to move the organism along. What has happened is that the turnover of focal contacts, and their integration into the bulk of the cell, have been disrupted, and so they can no longer assist in locomotion. This piece of evidence, along with the observations from interference microscopy and electron microscopy, makes it extremely unlikely that focal contacts are artefacts.

1.4 How cells evolved

My next dispute over the meaning of evidence concerns mutualism and the origin of the eukaryotic cell: a cell containing a nucleus, cytoskeleton and cellular organelles. Mutualism is the beneficial living-together of organisms. The general phenomenon of evolved relations between different species now has the umbrella term 'symbiosis', and includes both parasitism (where one organism is harmed) and commensalism (where neither harm nor benefit seems to occur to either species). A common example of a mutualist relation is the one formed between the shark and the pilot fish that clears the shark's gills of parasites. The shark protects its guests from predators and in return gains relative freedom from parasites. A more intimate example is the lichen, which is a long-term association between algae and fungi. In the case of the lichen, and of the shark, the association is not itself inherited, but there are examples where a mutualist relation is passed on to the next generation. The plant *Psychotria bacteriophia* has an association with nitrogen-fixing bacteria. These bacteria are not only in the roots of the plant, they are deposited in the seeds before reproduction. Thus, the new generation is launched not only with a complement of parental DNA, but also with some helpful bacteria. Mutualist relations can be very complex, involving more than two partners. Termites, which cannot on their own digest their woody diet, rely on other organisms living in their guts to provide the proper enzymes. One of these guests is the protoctist *Myxotricha*. The protoctist benefits from shelter and a good supply of food, and digests enough wood to provide a good source of nutrition to the termite. In this case, however, even the protoctist is an association: it harbours on its outer coat colonies of motile, whip-like bacteria called spirochaetes, whose energetic undulations propel the protoctist forwards. This seems like an extraordinarily intimate cohabitation, and is greatly celebrated by biologists interested in mutualism. It was these enthusiasts who in the 1970s put forward and defended a theory claiming that eukaryotic cells were themselves a result of a mutualist association between two (or more) organisms; this was called the Serial Endosymbiosis Theory (SET).

Like most good ideas, SET was not completely new. At the beginning of the twentieth century, cell biologists had became interested in mitochondria and chloroplasts. Previously, though visible, these organelles had been somewhat neglected. As described earlier in the chapter, the main focus of cell research had been the chromosomes, the nucleus and the mitotic process. It turned out that the staining techniques developed for showing up nuclear detail destroyed the mitochondria. This was simply a feature of the alcohol and acetic acid used to penetrate through to nuclei: in the process, the mitochondria fell apart. However, now that the chromosomes had been understood as involved in inheritance, some attention was focused on the rod-like mitochondria anchored in the cytoplasm. In the decade 1910–19 at least 500 papers were written on mitochondria, mostly in Germany and France. One of the observations reported was that mitochondria closely resembled bacteria. Moreover, mitochondria reproduced themselves separately and independently of cell division. A few maverick biologists wondered whether they might actually be bacteria. Among the first to develop a theory about the origins of mitochondria was the American biologist J.E. Wallin, during the 1920s. He suggested that mitochondria originated as bacteria, ingested by cells a long time ago, but now domesticated as a permanent, inherited feature. He also speculated that such acts of symbiosis generated genetic diversity, through the mixing of chunks of genetic material from different species. Wallin's ideas did not make much headway. This was the time of heroic fights against the microbial menace, with health construed simply as asepsis: the absence of microbes. Bacteria, then more than now, had a bad press. They were responsible for infectious disease, and the new knowledge of the menace of microbes did not sit comfortably with the thought that long ago bacteria had converted a primitive cell into the modern eukaryote. This was also the time that the modern theory of Darwinism was being refined by mathematicians and geneticists, such as J.B.S. Haldane, Sewall Wright, R.A. Fisher and Dobzhansky. They were working on ideas of gene mutation, gene frequency and fitness, and argued that gene selection could lead to evolution, to diversity. There was no need in their models for sudden boosts of diversity by the wholesale shift of genomic material from one organism to another. Slow mutation, and slow selection, would be enough.

Mutualism remained an obscure research programme throughout the 1950s and 1960s (parasitism, by contrast, has an obvious agricultural and medical importance). It is a field that attracts those who are suspicious of the modern scientific obsession: specialisation. People who study mutualism like to be involved in many areas at once. To study associations between animals and plants, or microbes and eukaryotes, you may need to be expert in areas ranging from ecology and geology to cytology, genetics and biochemistry. If you are interested in mutualism you will be reasonably welcome in half a dozen university departments, but completely at home in none. There is another reason why mutualism has attractions: it discusses the benefits of co-operation, and suggests that mutual aid is natural. One scholar has even noted that the first major conference on symbiosis, in April 1963, took place at the time of one of the Cold War's major alarms, the Cuban missile crisis.[7]

Although mutualism may not be the best-funded branch of biology, mitochondria have been researched intensively. Electron micrographs reveal a curling double membrane, with great partitions jutting into the inner space. These walls are spangled with proteins, known to be the site of the metabolic pathways of oxidative phosphorylation. With the new biomolecular technology developing in the 1960s, mitochondria and chloroplasts have been shown not only as reproducing themselves, but as having their own DNA. Biologists see all living things as showing one fundamental divide: organisms are made either of prokaryotic cells or eukaryotic cells, with no intermediate state known. Prokaryotic cells (bacteria and blue-green algae) are small, and without nucleus and organelles. Eukaryotic cells are large, with many organelles taking charge of aspects of the cell's processes. This is the greatest evolutionary discontinuity on earth. The biochemical and genetic research carried out on bacteria (for example on respiration, or gene expression), shows basic molecular similarities between prokaryotes and eukaryotes, but the gross cellular organisation of the two types is completely different. The question is unavoidable: how did eukaryotic cells evolve? There are two possible answers: either the prokaryotic cell developed on its own a more complex cytoplasm, complete with organelles; or the eukaryotic cell formed, suddenly, when two or more prokaryotes merged, to produce an entirely new entity. This is the

[7] Jan Sapp's book *Evolution by Association: a History of Symbiosis* (New York: Oxford University Press, 1994) is a detailed history of this research field. SET forms an excellent case study of a controversy in biology because it took well-known data and interpreted them in a novel fashion. Once again we see that when biological ideas are in conflict, the factors that eventually 'close' the argument are very diverse, and certainly extend beyond data from fresh experiments.

thesis of SET, and its main proponent is the American microbiologist Lynn Margulis.

You can get a sense of the ambition and interest of Margulis' work simply by looking at the titles of three of her publications: 'On the Origin of Mitosing Cells' (1967), *Origin of Eukaryotic Cells* (1970) and *Symbiosis in Cell Evolution: Life and its Environment on the Early Earth* (1981).[8] These are not works dealing with insignificant matters, for in addition to explaining the origin of mitochondria and chloroplasts, Margulis has also looked at other organelles in the eukaryotic gallery, and argues that centrioles and flagella also derive from prokaryotes. Her evidence for this latter idea is drawn from the example noted earlier: the protoctist *Myxotricha*, which relies on the whiplash motion of *Spriochaete* bacteria living on its skin. Spirochaetes have a similar structure to flagella; perhaps, Margulis wonders, the flagella that are so familiar a part of protoctist (and therefore eukaryotic) structure, were once free-living bacteria flailing about in the water. I will analyse these ideas as an example of scientific controversy, for in the early 1970s, scientific opinion was completely divided on the plausibility of this theory. Over a number of years the same sets of data were interpreted in completely divergent ways by the two sides in the debate. Like our eighteenth-century example from embryology (preformationism and epigenesis), this later debate is an example of how one set of results can have two or more explanations.

The prokaryotic cells arose between 4.5 and 2.7 billion years ago; the eukaryotic cells arose between 0.5 and 1.0 billion years ago. There are no intermediate forms in nature, nor in the fossil record, and there is no straightforward evidence of how the eukaryotes emerged. Margulis suggested, essentially, that several symbiotic events led from one form to the other, and that swallowing without digesting – endosymbiosis – is the best explanation. Put very briefly, Margulis imagined an anaerobic bacterium ingesting an aerobic bacterium (i.e. a protomitochondrion). In this thought-experiment, the next step is that the ingested microbe remains alive, and starts to metabolise the sugars inside the host. The guest takes advantage of the supply of food, but is leaky enough to offer a supply of ATP (adenosine triphosphate) to the host.

According to SET, all these events took place when the earth's atmosphere was accumulating oxygen: natural selection would, therefore, support a symbiotic relation that allowed a cell not only to survive oxygen,

---

[8] *Journal of Theoretical Biology*, **14**, 225–74; New Haven, Conn.: Yale University Press; San Francisco, Calif.: W.H. Freeman.

but also to use it. The partnership would work best if the host was active in searching for food, for example, by becoming amoeboid; but according to Margulis, another symbiotic association now occurred, with the attaching of flagellar-like bacteria to the outer surface of the symbiotic cell. The inner guests in time became the mitochondria, while the outer hangers-on became incorporated as flagella (or cilia). Moreover, according to the thought-experiment, another role was given to the flagellar guests. The machinery that sits at the bottom of any flagellum – the basal body – is very similar to the microtubular apparatus that controls mitosis, the centriole. Might it be that mitosis, another diagnostic feature of eukaryotic cells, evolved as a direct result of the simple ingestion of a microbe? Writing in the *Scientific American* in August 1971, Margulis made plain the importance of such a step: 'Without mitosis there could be no meiosis . . . no complex multicellular organisms and no natural selection along Mendelian genetic lines'.[9] Finally, another symbiosis might have taken place, when free-living photosynthetic blue green algae were taken in by the new heterotrophic eukaryotes. This latest association meant yet another kind of cell: a photosynthesising, aerobic, mitotic eukaryote.

It is easy to see why SET startled modern biologists. The theory is disturbing in several ways. Perhaps the most fundamental problem is its relationship with Darwinism. For according to SET, some of nature's biggest innovations in design have occurred not through the gradual process of mutation and selection, but through the gross swallowing of one organism by another. SET is not incompatible with Darwinism: were these associations bad for survival, they would not have lasted long. However, by focusing on events that occur between whole organisms of different species (and their associated genomes), the theory implies that gene mutation is not the only way that biological innovation can occur.

No laboratory has ever managed to repeat the events of symbiosis, for example, by growing mitochondria outside a cell, and subsequently reintegrating them into cytoplasm. This is an impossibility because though mitochondria have some DNA, enough has been integrated into the nuclear genome to make the nucleus a necessary element in mitochondrial reproduction. So can the Margulis theory be tested? Like many theories in biology (especially those that deal with evolution), SET is historical; the changes it discusses happened a long time ago. Yet while direct experimentation might be impossible, there are many observations from

[9] 'Symbiosis and evolution', *Scientific American*, **225**, 48–57.

different fields of biology that have been put to service in the debate. Amongst the most recent evidence has been that from gene sequencing technology. It was these data, gathered first in the early 1970s, that were used both to support and to undermine the theory. This interesting contradiction can be seen best in the data concerning the DNA found in mitochondria. Mitochondrial DNA is more similar to prokaryote DNA than it is to nuclear DNA. You might take this as clear evidence that the mitochondria do indeed have a prokaryotic origin: they retain, in their DNA, a memory of their prokaryotic past. When the data were first announced though, another interpretation was put forward, claiming the findings as evidence for the theory that mitochondria arose within the cell. For if the first eukaryotes formed when prokaryotes actually built their own mitochondria, rather than ingested them, it might have been easier for eukaryotic cellular organisation if the mitochondria had their own DNA, so reducing the responsibilities of the nucleus. Out in the cytoplasm, away from the intensely busy and selective environment of the nucleus, mitochondrial DNA might have evolved rather more slowly: a country, rather than a city life. After enough time, one would expect the mitochondrial DNA to 'fall behind' the nuclear DNA, but its primitivism relative to the nucleus would not then be because of separate origin, but because of separate rates of evolution.

It is often said by scientists and philosophers that a good theory is a simple one, and that it is a good principle to prefer a simple theory to a complex one. The principle is named Ockham's Razor, after William of Ockham, a fourteenth-century philosopher. Ockham's razor is applied when a choice must be made between two theories that differ from each other, but claim to explain the same set of data. In the early 1970s, with the SET being debated and similar evidence being used both to support and condemn the theory, Ockam's Razor was used in an attempt to end the argument. Thus, if symbiosis was the simplest explanation of how eukaryotic cells evolved, then it should be adopted on those grounds alone; but this is not what happened. Instead, for a while, there was a subsidiary dispute over which theory was indeed the simplest. In particular, the argument concerned the mitochondrial DNA. As suggested earlier, some of this material can be found in the nucleus. Opponents of SET said that the presence of mitochondrial DNA in the nucleus stood in their favour. For now SET had not only to suggest a mechanism for the ingestion of bacterial protomitochondria, it also had to explain how DNA could move from a mitochondrion to a nucleus. If mitochondria had evolved directly within the cell, from the nucleus, then the existence of mitochondrion-controlling DNA in the nucleus might – simply – be expected. By 1974, this example of the application of Ockam's Razor was being questioned: the technology of genetic engineering was being developed, and genes were being taken from bacteria and spliced into higher plants – gene movement was no longer 'a complicating factor'. If bacterial aerobes were ingested, it would be reasonable to expect some incorporation of their genome into the nucleus.

Here, then, is another example of two theories, and one set of data, and there seemed to be no independent evidence to choose definitively between them. So other considerations came to the fore, and no doubt there was a role for prejudice against a startling, slightly un-Darwinian idea. Ockham's Razor was widely cited at this time, but also, interestingly, disputed. The proponents of SET challenged their opponents: why are simple theories aways best, why must Ockham's preference for simplicity rule over the whole of biology, and is this itself not just a prejudice, a value? Margulis herself argued that scientific theories should be judged not only on their simplicity, but also on their plausibility and their ability to make use of different fields of expertise. She suggested that Ockam's Razor might not be such a useful arbitrator for interdisciplinary fields where many types of evidence and levels of explanation are being deployed at the same time. This was certainly the case with SET. It is routine for those interested in symbiosis to maintain a good knowledge of geology, atmospheric chemistry, oceanography and palaeontology, not to mention molecular biology, genetics and evolutionary theory. SET might be admired, not so much for its simplicity, as for another characteristic sometimes favoured by scientists: heuristic value, or 'fruitfulness'. Moreover, according to Margulis, the theory was a good scientific one in that it made predictions that could be tested: it might not be possible to travel back in time a billion years, but if SET is true, then a prokaryotic fingerprint might be expected in a whole range of eukaryotic mechanisms, ranging from mitosis to photosynthesis.

In the end, it was not the merits of Ockham's Razor that settled the dispute over SET. Instead, new and independent evidence arrived, making endosymbiosis a more plausible piece of science than its alternative, autogenetic, theory. Nucleotide sequence analysis of different groups of bacteria produced a distinction between two fundamentally different forms: the Archaebacteria and the Eubacteria. Clearly, if the DNA from a eukaryotic mitochondrion could be traced back to one of these groups, and the nuclear DNA to another, this would be convincing evidence of separate bacterial origins. The tests showed that this is true: mitochondrial DNA and nuclear DNA have distinct ancestors. So at the beginning of the twenty-first century, the basic controversy over endosymbiosis is now considered largely closed. The dispute has moved on: there are questions over the possible endosymbiotic origins of the nucleus, the centriole and the flagellum, and on these issues, symbiosis is still fought over. The most disputed idea of all, however, is that the symbiotic transfer of genes from one species to another has been an important feature throughout evolution, not only for the formation of the eukaryotic cell. If symbiosis ever became accepted as a motor for evolution in the metazoa, then the textbooks must all be rewritten.

1.5 The philosophy of science

The theme of this chapter is that science does not deliver certainties. My examples, both historical and contemporary, show how the ideas of science are not only debated and disputed by the professionals, but are also influenced by other ideas from far outside the laboratory. In particular, I have suggested that in a dispute between two contrasting theories, the biological evidence available may not be enough to choose between the two. In my example from the eighteenth century concerning development, data from the microscope could not adjudicate between preformationism and epigenesis; it was ideas from philosophy that first tipped the balance towards epigenesis. Before that, other philosophical ideas had favoured preformationism, since in that theory, the unfolding of miniaturised forms present since the beginning of time meshed well with the idea of Newton's universe: mechanical, but designed by God.

For some people, it comes as a surprise that the career of scientific ideas could be influenced by philosophical, social or political winds. If you believe that science is a laboratory matter, where strict ideas about experiments rule out any influence from the investigator's personality, religion or political beliefs, then you will be very sceptical about the idea that scientists need ever take seriously disciplines like the history or philosophy of science. For if the interpretation of scientific experiments is an activity wholly independent of the wider world of ideas within culture, then certainly any philosopher who tries to show how society influences science can be ignored completely. Needless to say, that is not my view; this whole book is largely an attempt to put your own study of biology into a wider context.

In order to understand the development of biological ideas, and even of the facts that we most take for granted, we have to be prepared to accept the possibility that progress in science is significantly influenced by a variety of social factors. Philosophers label such factors 'non-epistemic', to show that they are distinct from the traditional descriptions of knowledge gathering in science (the descriptions of 'epistemology'). There is a long list of these non-epistemic factors, and you will have views on which ones you consider likely to influence the course of scientific research. They include, in no particular order: funding, gender of the researcher, seniority of the researcher, ambitiousness of the researcher, status of the laboratory, attitude of the public to the importance of the research, and society's wider intellectual attitude. Clearly, the list could be much longer. Equally clearly, there is something controversial here. Are philosophers and historians suggesting that a scientist is no more protected from the vagaries of fashion than is a Hollywood film-maker? It would be a profoundly radical step to suggest that experimental results have nothing to do with the make-up of nature, and everything to do with whim.

In the last 30 years historians and philosophers have worked out a view of science where society has an important role in influencing the work of scientists. This work has frequently been controversial and is interpreted by some commentators as suggesting that successful science is nothing more than a list of fashionable ideas, tied in to a set of social beliefs, rather than determined by the actualities of nature. You may sometimes find that philosophers who study the role of society in shaping science are caricatured as 'anti-science'.[10] In this final section, however, I want to ignore these contemporary concerns, and focus instead on the strand of analysis that attempted to show how science is separable from wider social forces. In particular, at the risk of implying that the great days of philosophy of science are now over, I want to consider two philosophers whose main

---

[10] The academic discipline that studies the immersion of science in society is called 'Science Studies', and is sometimes viewed with suspicion by those scientists and philosophers who do not like to see non-scientists commenting on scientific research. If you want a taste of what is meant by 'academic fighting', a good place to start your search would be in the book *Higher Superstition: the Academic Left and Its Quarrels with Science* (Baltimore, Md.: Johns Hopkins University Press, 1994). The authors, Paul Gross and Norman Levitt strike out in swashbuckling style against feminists, 'relativists' and other science studies scholars. A counterblast is the revealingly titled *Science Wars* (Durham, N.C.: Duke University Press, 1996). Inevitably you will come across the 'Sokal Hoax', a gossipy and bad-tempered interlude where the physicist Alan Sokal attempted to portray the science studies community as poor scholars by hoodwinking them into accepting a hoax paper for the humanities journal *Social Text*.

work was completed by 1970, and who died in the 1990s. They are Karl Popper and Thomas Kuhn. Although Kuhn has been interpreted as the more modernising philosopher, and is usually seen as the more influential on contemporary intellectual life, they both have in common a desire to 'show how science works'. In other words, they were completely unafraid of wading into the murky waters of science, and explaining to all who would listen exactly what were the rules of good science. Unlike most contemporary science studies scholars, they were prepared to preach; they were good writers too. I highlight their work partly because of their ambitious attempts to specify the method of science, and partly because their writing is thought-provoking and accessible.

Karl Popper (1902–1994)

Popper wanted to be clear about the distinction between science and non-science. He called this the 'problem of demarcation'. It was not that he considered anything that was not scientific to be useless, but the reverse, for Popper himself was very musical. However, he did think that it was important to be able to tell the difference, and it seems as though his emphasis on this was caused by an irritation with those political theories (e.g. communism) and psychological theories (e.g. psychoanalysis) that during Popper's intellectually formative years in the 1930s were routinely claimed by their respective adherents as 'scientific'. Popper did not want to say that psychoanalysis was useless; he just wanted to say that it was unscientific. He pondered as follows. No proposal about nature can be proved true. Unlike mathematics, or logic, where proofs can be arrived at by the method of deduction, proofs about nature simply do not occur. It is easy to see why. For example every eukaryotic cell membrane ever studied has a phospholipid structure variously penetrated by proteins, and we have got used to saying that 'all eukaryotic cell membranes have such-and-such a structure'. Yet our justification for that 'all' is, at least in terms of logic, quite flimsy. Might we not come across a cell, some time in the future, that has an entirely different kind of membrane? It seems unlikely, agreed, but it could happen: why not? The past is not always a good guide to the future. It was the Scottish philosopher David Hume (1711–76) who first pointed out the problem of generalising from repeated examples to a scientific proof; the issue came to be known as 'the problem of induction', and has been an issue for philosophers ever since.

The problem of induction is not something a cell biologist is going to worry about. Popper, however, did worry. Being a philosopher of a certain type, he was concerned to show that even though science could not have the certainty of mathematics, it nevertheless was a different kind of thing from other uncertain fields of enquiry, such as political theory and economics, or psychoanalysis. He developed the principle of falsification to drive the difference home. Popper's so-called hypothetico-deductive method, which makes use of the idea of falsification, is easy to grasp – a valuable characteristic of this particular philosopher. It goes like this: initially, the scientist, perhaps contemplating something seen down the microscope, creates a hypothesis. The hypothesis is a scientific one if it makes predictions that can be tested, but the aim is not simply to test the prediction to make sure that it happens. On the contrary, says Popper, the emphasis must be on testing it to destruction. For if the prediction (derived logically from the hypothesis) is not met, then, logically, we know the hypothesis is certainly false. Thanks to Hume's problem of induction, we cannot derive the same certainty from a prediction that works. All we can say then is that the hypothesis is supported. Thus there is an asymmetry, with the refutation of a theory being a more solid base to science than its support. In sum, science is a thin veneer of surviving, testable conjectures, any or all of which may someday topple.

Scientific experiments are those that can go in one of two ways. If the experiment shows that the hypothesis is wrong, then the hypothesis is thrown out, and the work must start again. If the experiment does not throw out – falsify – the hypothesis, then it is supported (Popper used the word corroborated), but not proved. According to Popper, two simple messages emerge from this idea. Firstly, only those hypotheses which bear within them the possibility of falsification, the seeds of their destruction, can be scientific. Secondly, it is the mark of good scientists that they are bent on submitting their theories to stringent testing, throwing them out as soon as an experiment declares against them. The progress of science is similarly easy to define. It is simply the survival of hypotheses during trial-by-experiment, with the replacement of poor hypotheses by better and better survivors. On this account, textbooks are populated by the most hardened of survivors, while research laboratories are far more tentative, and might at any time have to jettison the fragile theories currently under test. Still, some of the theories in those laboratories do survive, are seen to be reliable, get incorporated into a stable body of knowledge, and, in time, make the textbooks.

Popper's ideas have collected a number of criticisms. It does not seem likely that scientists in the real world throw out their theories just because of some counter-evidence. A theory may make a wrong prediction not because it is completely false, but because of some minor problem in its detail. It would be better, therefore, to persevere with the theory and fix the problem. It is in a scientist's interest to build-up the solidity of a theory, and persuade others of its merit. Constantly throwing doubt on the value of ones work is not a proven career strategy. Moreover, the actual ambition of Popper in explaining how science works now seems dated. Is it likely that one method of science can be described as applicable to all sciences, from astronomy to zoology? In the mid-twentieth century, philosophers pushed in this direction; 50 years later it is far from clear that the project was successful.

Thomas Kuhn (1922–1996)

Karl Popper's approach fits well with the popular idea of the lonely scientist struggling in the laboratory, determined to find the truth and unfailingly adhering to the highest norms of honesty. You may have come across this image of science in films and books about Louis Pasteur and Marie Curie. With its atmosphere of honesty and endeavour the image is in some ways flattering, and no doubt represents some part of all our scientists. Yet there is a disturbing and malevolent aspect that can be superimposed on the Popperian vision: the idea of scientist as a loner, uncommunicative and unemotional, fixated on the overriding importance of the work.

Perhaps it is reflections like these that account for so much of the popularity of Thomas Kuhn. Like Popper, Kuhn writes comparatively clearly. The distinctive aspect of Kuhn's work is that he emphasises the idea that science is a community enterprise. Kuhn evokes a vision of science where groups of people work together on experiments, sharing ideas and conforming to traditions about which scientific ideas are worthwhile. Kuhn suggests that decisions on scientific ideas are not dictated simply by the results of experiments (as is the case in Popper's vision) but are vitally affected by the all-too-human tendencies to conform to the beliefs and desires of the group, or even of wider society.

Kuhn said that science does not advance by the continual replacement of falsified theories by better alternatives. On the contrary, science, for the most part, avoids novelty. In what Kuhn calls 'normal science', scientists within a particular area (for example cell biology) simply do mundane work, bolstering the particular theories then in vogue, flattering the professors, and writing very standard, yet career-enhancing, papers; no one is trying to upset any theory or any person. In Kuhn's view, rogue ideas and

novel hypotheses are viewed with extreme suspicion and distaste, and scientists who persist in pushing views outside the mainstream are likely to
be punished. Their papers are rejected by all the journals, they cannot get
secure employment in the university, and they acquire a reputation as a
'maverick'. Kuhn used a word to describe this pervasive and dreary group
agreement about what is true and what is interesting to study. He called it
a 'paradigm'. There has been plenty of argument over whether this term
means very much, and in particular over whether there can be said to be a
single paradigm governing, say, cell biology or evolutionary studies. However, for Kuhn, as described in his best-selling book *The Structure of Scientific
Revolutions*,[11] paradigms absolutely determine what a scientist can do and
believe.

Yet over time, in spite of the suffocating conformity imposed by the
paradigm, odd results accumulate and cause problems. Anomalies conflict with what had been thought to be obviously true; they can no longer
be ignored. The whole structure of the paradigm begins to shake; conflict breaks out and splinter groups form. As a result, perhaps through
the visionary thinking of a great mind – Charles Darwin for example – a
massive reworking of ideas takes place. According to Kuhn, this happens
sufficiently quickly to be aptly called a 'revolution'. The term 'revolution'
is suitable not simply because a change occurs rapidly, but because the
change is great. Indeed, as you have probably realised, the result of the
revolution is a change in the paradigm. What was distrusted and avoided
before is now routine and ordinary, the stuff of a new period of 'normal
science'. Kuhn's neat formula has proved extremely influential, and popular outside science as well. You may come across the phrase 'paradigm
shift' as a shorthand way of describing some radical change of thinking
within a person or a community. Kuhn's work is usually interpreted as encouraging sociological and political analysis of how science meshes with
society, but careful readings of Kuhn suggest that he was mostly concerned with the way in which the community of scientists worked, and
was not arguing that the wider society has any influence in how science
progresses.

In exchanges between the two men, Popper claimed that normal science was 'bad science', because it suggested that scientists were content to live with, and conform to, a status quo. Kuhn retorted that history shows that revolutions do happen, and that it is through these that

---

[11] Chicago, Ill.: University of Chicago Press, 1962.

science progresses. What evidence is there that revolutions have occurred within science? Within biology, the acceptance of the Darwinian theory of evolution has sometimes been cited as an example of a Kuhnian revolution. Loosely speaking, before Darwin published his book *The Origin of Species* in 1859 (see Chapter 3) the official view within scientific, ecclesiastical and political circles was that species were fixed and had not evolved. However, for many years strains and stresses had been appearing in this 'paradigm', as ever more fossils were discovered (suggesting that some organisms had become extinct while others had arisen) and as political changes, stemming most dramatically from the French Revolution, suggested that fixed classes in general, and the aristocracy in particular, were not an immutable part of nature, but could be challenged and indeed changed. It was Darwin's biological genius to take all the scientific data that could support his own view that evolution of organisms had taken place, rearrange it as evidence for his view, and present a whole new way of ordering nature. This work took place over 20 years and more, but was made public in a day. After Darwin it was commonplace to accept that organisms had evolved. However, such a description of Darwinism as a paradigm shift may be too simplistic. Was the change instant, did it straight away affect the whole of biology, and were Darwin's ideas completely incompatible with his predecessors? If Thomas Kuhn is correct, we would expect the answer 'yes' to all these questions. Scholarship into the history of evolutionary theory produces a more complex, uneven picture.

This chapter has explored the development of ideas, big and less big, in cell biology. The stories told do not fit completely with Kuhn's theories, or with those of Popper. The desire to find only a 'method of science', one that will fit all examples, may itself be a mistake. Science is diverse; even within biology there is a multitude of ways of working. However, while I have not tried to capture some essence of what it is to be a biologist, I have emphasised two points. One is simply that nothing in biology is certain. Another is this: nothing simple lies behind the many facts and figures that form so much of a biology student's diet. Those facts have not sprung, fully formed, simply from clever experiments by clever people: facts are ideas, and have histories and ancestors. Seen like this, biology becomes a subject with links to all of life, rather than a subject whose mastery involves only the learning of nature's lists.

2

Reductionism

2.1 An introduction

Biology is the study of living things. Its domain is life. Biologists inquire into the way things crawl or jump; they study how one generation makes the next, or how animals breathe and find their food. In short, biology asks questions about organisms. It explains how living things grow, reproduce and die.

Through this image of biology, we see the planet earth as populated by millions of kinds of living things, purposefully going about their business, with biologists meanwhile trying to fathom why they do the things they do. It is the biology of wildlife television programmes in the tradition of David Attenborough, of whales and lions, orchids and Venus flytraps, rock pools and mountain eyries. If for school students biology sometimes shines over physics and chemistry it is because the subject is about something 'real' – the human body, or animals, or plants. The idea lasts: many working biologists describe their initiating experience as a powerful encounter with living things close-up, with insects or Californian redwoods or sea anemones. No doubt these formative influences run deep in most biologists, so it is interesting to investigate how far the reality of modern biology matches up to our first hopes of the subject. A casual look at any general textbook, or course synopsis, provokes the feeling that biology is not, after all, about living things. Instead of animals and plants – or fungi or protoctists or prokaryotes – we see molecules, organelles and enzyme pathways. For most biologists, the living thing is understood by investigating the not-alive. To put the matter plainly, modern biology, while keeping half an eye on whole organisms, is mostly interested in the very, very small. The guiding principle for modern biology is this: to understand the organism you must understand the components it is made of. As a result, the modern biologist is likely to be much more comfortable with the molecules of an animal than with its migration.

Let us label this philosophy straight away: it is called reductionism, and can be defined as the belief that the best and most reliable explanations in science depend on a knowledge of components, and are expressed in terms of those components. This chapter will review some of the arguments about reductionism. Amongst biologists there is a debate over reductionism, and sometimes the arguments get quite loud. This is because reductionism is such a powerful trend in biology, and like all successful trends, squeezes out other ways of seeing things; that could be a significant loss. Some biologists and philosophers worry that the life sciences will produce a distorted image of nature if the emphasis is all the time on the examination of molecules. Could important lines of research wither, simply because they are not obsessed with molecules?

These arguments are important because they expose the way that biologists give priority to a certain class of problems, and a certain class of solutions. For example, since the 1950s, the most intensely studied class of problems has been those concerning the molecular make-up of organisms, and the most celebrated class of solutions has been those that involve genes. If molecular biology is a product of reductionism, then we must admit that this particular way of asking questions is very successful. The Human Genome Project is an obvious mark of the power of reductionism, but the limits are also significant. The drive to itemise each and every component generates gigantic lists of facts, vast terminologies, and correspondingly daunting textbooks. How well do these lists help us to understand the life of organisms, including humans? The problem of reductionism is that it casts a spell. Biological solutions remain trapped at the level of the molecule, always on the search for interesting mechanisms, but forgetful of the life of the whole organism. This can be seen with the Human Genome Project. Once, necessarily, it was an obsessive campaign to list nucleotides and to map genes. Now it is trying to scramble back up the scale towards the whole organism – hence the field of proteomics, which is an attempt to relate human genes to human proteins.

Meanwhile, other biological fields stay out of the limelight. These tend to be the disciplines where molecular techniques do not dominate. Ecology, evolutionary studies, systematics, taxonomy and ethology are all examples of disciplines that do important work, and sometimes make use of molecular techniques, but are not overtly reductionist in basic philosophy. These disciplines are important in biology, but they make much less of a splash than molecular research programmes. They are less expensive and less obviously related to human health, and because they do not attract much interest from the biotechnology and pharmacology industries, they do not have press offices and public relations professionals trumpeting their successes, and promising that there is more to come.

Before I attempt to analyse reductionist philosophy, one point must be made plain: its success cannot be disputed. The following list of some well-known types of biological components illustrates the point: deoxyribonucleic acid, genes, proteins, membranes, Golgi bodies, nuclei, cells, tissues, organs. All of these components (or classes of components) are further reducible to smaller elements. The value of these fragmentations is various. By having a labelled and defined component, for example a Golgi body, different biologists can talk to each other in the certainty of knowing that they are discussing the same item. Another important aspect is the fact that very often the same component occurs throughout nature, and is not restricted to a single species, organism, tissue or cell. For example the genetic code was first understood through research on microbes, but is universal to all living things, including of course *Homo sapiens*. Also, time after time, reductionism has solved biological puzzles. Think of the vertebrate muscle – how does it contract? A reductionist analysis provides a good part of the answer, with the myofilaments sliding past each other, thanks to the movement of strings of actin and myosin molecules. Very importantly, these reductionist successes are important because they do more than 'increase our understanding of nature'. Rather than just amass scientific detail, reductionist science suggests ways in which we can interfere with those components, in the most delicate, yet effective way. Modern medicine is based on reductionist biology, as is the genetic modification of crops, and reductionism suggests many lines of research for the future. In the field of genetics, for example, the great reductionist prize of discovering the structure of DNA is to be used in the medical process of gene therapy. Knowledge of the genetic code will be supplemented by a knowledge of how to chop out, disable, or replace strips of DNA that we as a society consider harmful.

Scientists, like philosophers, are quite choosy about the questions they aim to answer. They will only go to work when the question – perhaps we can call it a hypothesis – is tightly defined, capable of being tested, makes clear predictions. So what might be an obvious question for a biologist – 'What is life?' – is likely to be approached as a question of chemistry. It is the chemical constituents of organic life that will, thanks to the reductionist, dominate our questions into the basis of life. However, there is something unconvincing about the idea that life is nothing but chemistry. Molecules may be complex but they are not alive. A commonplace saying, often heard when people debate this problem, runs like this: 'The whole is more than the sum of its parts.' The suggestion is clear: an organism is more than a collection of molecules, or a collection of cells, or even a collection of organs. The person who considered life as simply molecules in motion, we in turn would consider too coldly scientific, or even wrong. Yet it is hard to pin down, to label as a thing, what it is that makes a living organism more than simply a big bunch of molecules, or a collection of organs. Those who are sceptical of the scope of reductionist biology use a terminology that stresses relationship. Concepts like 'organisation' or 'order' or 'complex system' are common in accounts that put a clear limit on the capabilities of reductionism. Ontogeny, the development of the individual from egg to adult, is one area where the powers of reductionism are contested: will knowledge of genes and other molecules do the trick, or will we need in addition another, more obscure tool kit – the concepts of complexity, of architectural constraint, and of systems? Yet though these terms certainly do the job of pulling our eyes away from lists of components, they remain vague and hard to define. That vagueness might be due to their unfamiliarity, since biology is inclined to look at the very small. A change of habit is needed to keep the big picture in mind.[1]

To sum up, the reductionist's approach to biology is to take organisms apart, identify the components, and study their connections. As argued above, the strategy is a proven success story. The anti-reductionist emphasis is always on the wider picture. Components are not ignored, but they are seen as part of massively complex networks. Those networks must be studied whole, for once disbanded into building blocks, the living

---

[1] The debate about the role of reductionism in biology has run for centuries. The controversy described in Chapter 1, concerning epigenesis and preformationism had as one of its themes a disagreement about reductionism. Caspar Friedrich Woolff (1733–94), sometimes described as the founder of embryology, refused to see biological development as simply the unfolding of miniaturised components. According to Shirley Roe, Wolff 'decried the total reductionism he saw in the application of mechanism to biology being made in his day'. (*Matter, Life and Generation: Eighteenth-Century Embryology and the Haller-Wolff debate*, Cambridge: Cambridge University Press, 1981). However, the allure of the sure foundations of physics and chemistry remained long after Woolf's death. Jon Turney's book, *Frankenstein's Footsteps: Science, Genetics and Popular Culture* (New Haven, Conn.: Yale University Press, 1998, p. 71) recalls an early twentieth-century biologist's reaction to a colleague: 'By brazenly parading his mechanistic animus, he did more than anyone to foster the belief that the most effective approach to biology and medical research is through physics and chemistry.'

thing is lost from the biological gaze, and the point of the biological enterprise has been forgotten. Anti-reductionism tends towards the interdisciplinary, reductionism towards the specialist. Very roughly, biochemistry, immunology and molecular biology are considered reductionist, while ecology, evolution and ethology are considered anti-reductionist. However, this classification is indeed very crude, for plenty of ecologists have a fine understanding of chemistry, and an immunologist may be studying the relation of environment to personal health. The point is that a judicious application of both approaches can be applied to any field in biology. Reductionism on its own distorts biology. Both its applications, and its personnel, become narrowed. Good biologists can see beyond their specialism, and make links with other fields. They like to put their work into the context of the life of organisms, or the health of society, or the health of the planet. A reluctance to look beyond the components of the system under study may be productive in the short run, but will eventually prove stultifying.

2.2 Examples of reductionist research

The neutrophil

The neutrophil is a macrophage. It ingests dead cells, bacteria and foreign antigenic material clumped by antibodies. A neutrophil exhibits, therefore, remarkable dynamic qualities. It can change shape dramatically, it can crawl, and it can respond to the presence of foreign bodies; but how does a neutrophil sense 'the enemy' and discern its orientation? Reductionist research strongly favours a chemical investigation. We already know something of the chemistry of the cytoskeleton and of the role of actin and actin-folding molecules in developing intracellular forces. We would expect the neutrophils to sense an invading pathogen chemically, in the same way that we smell bacon. We can imagine a range of questions that need to be answered. These include: what is the chemical sensed, where does the chemical come from, how does the neutrophil determine direction from exposure to this molecule, and how does the cytoskeleton of the neutrophil react once the molecule has been sensed?

Research has shown that neutrophils can detect, using receptors in the cell membrane, a particular N-formylated peptide, characteristically found in bacterial membranes. For the neutrophil, therefore, the peptide denotes enemy. However, research has also shown that the receptors are very sensitive to the concentration of peptide, so different receptors located on different areas of the neutrophils surface can detect slight variations. The tiny concentration gradient of peptide, set up along the length of a neutrophil when it is in the vicinity of a bacterium, stimulates the receptors to a greater or lesser degree. Thus, the neutrophil can compute the direction of the source of peptide. This information, in some way not understood, causes a range of effects in the cytoplasm of the neutrophil, which accordingly moves towards the bacterium.

Much more is known about what goes on inside cells than between cells. Presumably this is because inside a cell there is an abundance of structure to study, as well as chemical processes. However, when it comes to studying what goes on between cells, chemical signalling, and sometimes physical forces, become the dominating interest for the reductionist researcher. A knowledge of intercellular interactions is fundamental to diverse aspects of modern biology, including cancer research and embryonic development. Most fundamentally, a knowledge of intercellular relations will enable us to understand better how multicellular organisms arose in evolutionary history and how the component cells of such an organism cooperate within the common purpose. The reductionist approach to this rather complex problem is, amongst other things, to give priority to research into chemical signalling.

The slime mould

In the context of research into chemical signalling, the slime mould is particularly interesting. If we focus on one type, the cellular slime mould *Dictyostelium discoideum*, we see over the course of its lifetime an oscillation. Sometimes *Dictyostelium* is a free-living amoeboid protoctist, whose habitat is the woodland floor, and which moves slowly through the soil or humus, ingesting bacteria. Sometimes however, it forms colonies, congregating together with millions of others to form a slug up to 3 mm long. This slug is perhaps more than a colony, in the same way that a metazoan, for example an earthworm, is more than a collection of cells. For the slug moves over the ground, traversing obstacles and climbing logs. Indeed, the slug has a specialist activity: reproduction. After a period of time the slug sends up a long periscope-like shoot, on the summit of which lies a spore-producing pod. Having reached the air, spores waft away and, on landing some distance from the parent slug, germinate into free-living amoeboid protoctists.

The slime mould is an extraordinary creature. It raises questions about what we mean by metazoan, colony and organism: which of these three terms is most appropriate for the slug? Furthermore, of even greater relevance to our purposes in this discussion, the ability of individual cells to come together to produce a coherent macroscopic mass capable of doing something no cell on its own could do – sporulation – brings to mind the familiar slogan: 'The whole is more than the sum of its parts.'

Yet reductionist biology has in fact told us a great deal about slime moulds. Like the neutrophil, the slime mould has orientation. It seems to have a front and a back. Moreover, the ability of the slug to move in one direction, and then to marshal thousands of its member cells to take up position as a perpendicular stalk, with sporule held aloft, suggests a high degree of control. For the reductionist, the search much be for a framework of chemical signalling, both within the slug, and between the slug and the environment.[2]

These two examples illustrate how successful reductionist biology can be. You might even think, simply from the brief details given in this section, that neutrophils and slime mould have been in all essentials understood thanks to the probings of chemistry, and that for neutrophils and slime moulds the subject is now closed, with everything a scientist needs to know, known. By the end of the chapter however, with question marks having been raised about the scope of reductionist biology, you will be able to refer back to these examples, and imagine less reductionist, but perhaps equally fruitful, lines of inquiry.

2.3 The two faces of reductionism

We can distinguish two different aspects to reductionism: the pragmatic and the philosophical. Put like that, you can see I am suggesting the latter as a more fundamental belief, while the former is simply the get-through-the-day strategy of a busy biologist. If you are a biology student you will come across pragmatic reductionism every day of your studies. Philosophical reductionism is, however, a belief about the deep structure of the universe, and you are more likely to encounter it if you come across

---

[2] The US National Institutes of Health (NIH) have chosen *Dictyostelium* as a model organism for gene sequencing, as can be seen from an associated web site (http://www.nih.gov/science/models/d.discoideum/): 'The hereditary information is carried on six chromosomes with sizes ranging from 4 to 7 Mb resulting in a total of about 34 Mb of DNA, a multicopy 90 kb extrachromosomal element that harbors the rRNA genes, and the 55 kb mitochondrial genome. The estimated number of genes in the genome is 8,000 to 10,000 and many of the known genes show a high degree of sequence similarity to genes in vertebrate species.' It will be interesting to see how quickly this information is related to the details of the mould's lifestyle.

a philosopher, or a particularly philosophical scientist. You should be aware though, that this is not a completely clear-cut division, and I accentuate it only for the purpose of clarity.

For the pragmatic reductionist, component-centred biology is simply the most effective way of doing research. The history of science shows, as a matter of fact, that reductionism works. The task of analysing life at the cellular and molecular level has brought great rewards. When you turn your attention to components, the apparent differences between organisms begin to disappear. Cells – all cells – have much of their chemistry in common: you can do research on yeast, and hope to extrapolate your findings to humans. The pragmatic reductionist is attracted by the possibility of finding simple causal explanations for complex biological phenomena. By this I mean that the eventual explanation works by showing how one chemical event in the cell leads to another, and then another. These causal explanations show how the products of one reaction are the raw materials of another. The explanation relies on knowledge of chemical reactions, on the making and breaking of bonds, or even of quantum events at the subatomic level. Thus, the reductionist sees in the cellular expression of the genetic code a series of causal links leading from deoxyribonucleic acid (DNA) to manufactured enzyme, links best understood as a series of chemical reactions involving ribonucleic acid, ribosomes and so on. Another example is the citric acid cycle, a loop of chemical reactions dependent on each other.

Philosophical reductionism

The more radical brand of reductionism, philosophical reductionism, joins its pragmatic counterpart in believing that explanations at the chemical or physical level represent the best hope for biology. This is not only because of the enticements of apparently simple causal explanations, or because results come quicker, or because an understanding of biomolecular pathways can lead in turn to new technologies and new therapies. It is also because, for the philosophical reductionist, the only real things in the world are subatomic particles. Such particles, and their interactions, are the bedrock of reality. By this way of thinking, it is fine to wonder about cells, but if you want to get at the firmest foundations for your work, you must burrow down and search for atomic explanations of the cellular phenomena. Explanations based on the movement and behaviour of particles will be the best, the most fundamental, and the truest. These explanations are the worthiest because they are based on fundamental entities. Everything else one comes across in biology – molecules, cells, organisms, ecosystems – is simply a loosely knit association, hardly real at all.

A further goal for the philosophical reductionist is to unite the sciences. The current division of the sciences, which puts partitions between biology, chemistry and physics, depends on research being carried out at different levels: at the level of the organism (traditionally) for biology, at the level of the molecule for chemistry, and at the level of fundamental particles for physics. Philosophical reductionists find these divisions deeply unsatisfactory. For them there is only one science, the science of the atom, or, better still, the quark. If we can find the laws which govern the motion of atoms, and how they interact, then we can explain the phenomena of chemistry. Once those laws governing chemistry have been found, then we can explain the chemical processes going go inside the cell, and when that has been achieved, we can explain the behaviour of cells and how they interact. In the end, the complexities of organismal life can be linked back to the behaviour of atoms, and indeed explained by them. Effectively, biology and chemistry disappear, with all the phenomena and theories of these subjects amply covered by the laws of physics; thus science becomes united. In other words, if we can reduce everything to physics, we can get rid of artificial subjects like biology, or worse, sociology, and search for and settle on a few fundamental equations that once and for all will explain the universe: the one Theory of Everything. As the physicist Stephen Hawking put it, we will know 'the mind of God'.[3]

There are several problems with this philosophical reductionism. Firstly, in the real world, any attempt to explain all of biology by a description of atoms would quickly fail. When a mountain gorilla in a zoo stares through a plate glass window at a 3-year-old child, very little of what we know about the gorilla's atoms will help us find out what is in its mind, or in the child's mind. Nor can a knowledge of atoms tell us much about Leonardo da Vinci, football hooliganism, or an episode of the television programme 'Friends'. To put it another way, the interesting and important things that you might want to understand about football hooliganism are just as likely, or more likely, to be arrived at by understanding a range of social issues – from peer pressure and questions of self-esteem, to alcohol legislation, policing policy and stadium design. To make headway here, the philosophical reductionist, while admitting that there is

[3] *A Brief History of Time: From the Big Bang to Black Holes* (New York: Bantam, 1998), p. 193.

no current prospect of explaining elephant behaviour in terms of quarks, must nonetheless produce some rules showing how a less ambitious reduction, perhaps only from the biological to the chemical, might work. The major task for the reductionist is to show that nothing important, no essential insight or avenue of research, is lost when some aspect of animal or human behaviour is explained in terms of chemistry: when, in short, the sociological, psychological or biological is abandoned in favour of the chemical bond. Sometimes, criminal or aggressive behaviour is explained in terms of levels of a neurotransmitter in the brain, such as serotonin, or the male hormone testosterone. In cases like these, the reductionist has to show not only that serotonin is involved in some way, but also that we have no need for sociological theories to understand criminality: that abnormal blood chemistry fully explains abnormal behaviour.

2.4 The general problem of scientific explanation

In the preceding sections of this chapter, the word explanation is used as though it is a simple concept. Clearly a job of the scientist, perhaps the main job, is to find explanations for natural phenomena. Yet it is quite hard to describe what exactly we mean by explanation, beyond the obvious point that we expect an explanation to help us understand how something works, or why it happened. Generally speaking, people believe that a 'scientific explanation' is particularly valuable when it comes to accounting for natural phenomena ranging from lunar eclipses and desert storms to the behaviour of bees and the coloration of plants; but other sorts of explanation, that appear non-scientific, exist as well. There is nothing unusual in people looking for a non-scientific explanation for an event. If red pigment leaks out of a plaster saint's eyes in an Italian village, the Vatican is likely to take seriously the idea that this is the hand of God. Conversely, in other cases, a scientific explanation is not likely to take you very far: for example in explaining the kind of music you like, or the kind of job you want. Yet even if scientific explanations are not appropriate for everything, they certainly constitute a valued tool for understanding a great many things that happen, or do not happen, around us. Given the success of science over the last 300 years, it is not surprising that philosophers are always trying to summarise the precise nature of its structural foundation, namely, the scientific explanation.

One stereotype of scientists is that they are austere, passionless and logical. We do not have to subscribe to this view, but it gives us a hint that logic may be a valued component of the scientific explanation. Logic is a branch of philosophy that studies valid reasoning and argument. In logic, the deductive argument shows how one thing may be said to follow from, or be a consequence of, another. Deduction is obviously going to be an important part of a scientist's mental tool kit. It would be a bad sign if a scientist made contradictory statements within a paper, or drew an illogical conclusion from an argument within that paper. To show what I mean more clearly, here is a very simple deductive argument:

(a) All humans are mammals
(b) John is a human
Therefore
(c) John is a mammal

The first two lines, (a) and (b), are premises, like definitions. The third line follows deductively from the definitions. We can see that truth is conserved. Because the premises are true and the conclusion valid, the conclusion is also true. Yet a deductively valid conclusion does not have to be true:

(a) All humans are fish
(b) John is a human
Therefore
(c) John is a fish

This is a deductive argument, of exactly the same format as the first. It is logical but untrue, because premise (a) is incorrect. This gives us a clue to a feature of science. Scientists must respect logic, but they cannot rely solely on it. We know that an important task of scientists, probably the most important task, is to 'interrogate nature'. Scientists' statements include observations, often generated by experiment. These observations (for example of a particular mutation on a gene) involve a whole mix of operations: skill, machines, luck, knowledge and patience. Scientists, in other words, go beyond logic because they have to experiment with nature. They are creative people, looking for new ways of investigating and describing. Whereas logicians simply derive conclusions that are contained within the premises, and analyse the structure of arguments for signs of falsehood, scientists look outwards and use their skills to make something new: though they use language for describing nature, they are not simply investigating language.

Scientists, in other words, use more than logic; they also work with observations of nature. This is a complex mix, so it is not surprising that philosophers have not been very successful in giving a simple, widely applicable, description of the elements of scientific explanation. Their task is hard for the following reasons. Firstly, scientists conceive and plan their experiments in many different ways, and may not even be aware of how they get their ideas. Secondly, following from this, it is obviously hard to establish the relation between austere logic, and the creative, practical world of scientific experimentation. The result is that whenever one philosopher proclaims a general mechanism underpinning all satisfactory explanation, another begins to list the exceptions to the rule.

The earliest modern attempt to isolate the principles of scientific explanation took place in the 1940s. The American philosopher Carl Hempel wrote a classic essay called 'Studies in the logic of explanation'.[4] The paper describes a mechanism of explanation called 'the covering law model'. The central characteristic of the model is the central role given to scientific laws. The model tries to unite the rigour of logical argument (which should be part of science) with the observations and experimental results of the working scientist (which also should be part of science). A good example would the explanation of the colours of the rainbow. Its multicoloured appearance is to be explained by the laws of optics, and by the knowledge that, high up in the sky, sunshine is passing through a great many drops of water. The same procedure could be used to explain a lunar eclipse. In this case, the eclipse can be considered explained if the temporary blanking out of the moon can be shown to be a consequence of Newton's law of gravitational attraction and some knowledge of the speed and mass of the actual moon.

This is not the place to discuss in detail the reasons for the failings of the covering law model, though you can probably guess the outlines of the critique. The question is over how widely applicable Hempel's programme can be. Whereas it is not difficult to think of laws in physics and chemistry, in biology the task is harder. If by law we mean a timeless truth applicable without exception throughout the universe, then biology cannot have any laws. By definition, biology thinks about life on one planet:

[4] Carl G. Hempel with Paul Oppenheim, *Philosophy of Science*, 15 (1948), 135–75.

earth. As it happens, some philosophers question the absoluteness even of the fundamental laws of physics, doubting that they are exceptionless, and asking how we can describe as 'true', a law that holds only in certain circumstances.[5] Yet whatever the wobbliness of the laws of physics, it must be admitted that biology has even less claim to be law-like. Have you ever referred to a law when giving a biological explanation? – almost certainly not. Of course, the laws of physics and chemistry are relevant to the actions of the atoms and molecules within the cell, but even in molecular biology, we are not simply interested in the making and breaking of chemical bonds. It is the actions of molecules in building, perpetuating, or destroying biological mechanisms that interests us. We do not do molecular biology solely by referring to the rules of chemistry (valencies, bonds, forces), and we do not explain genetic phenomena by turning to the laws of chemistry. None of this denies that biology is a science. It is simply that biologists work with regularities that differ from the kind of regularities of physics and chemistry. So now we can see how the problem of explanation relates to reductionism. If biology explanations seem to be of a kind different to those of chemistry and physics, this might be taken as an obstacle to the easy reduction of biology to chemistry and physics. If biologists typically use explanations that are neither chemical nor physical, and philosophers have given up the attempt to define the 'best explanation', then, clearly, there must be something wrong with the idea that all discussions on evolution can in principle be reduced to discussions about atoms.

2.5 Explanation through causation

As suggested earlier, philosophers have not been united in thinking that laws must be the backbone of scientific explanations. At about the same time that Hempel put forward his ideas, the philosopher Michael Scriven argued that the explanation for a phenomenon must rest on a description of the causes that led to it; but just as some philosophers have problems with the concept of physical law, so do they with causation – the idea

---

[5] See for example Nancy Cartwright's book *How the Laws of Physics Lie* (Oxford: Oxford University Press, 1983). For a philosopher, the concept of scientific truth turns out to be a tricky one. One of the questions that may be asked is whether something in science is said to be true because it fits well with other established scientific ideas, or whether, going further, a theory or explanation is said to be true because it corresponds to the actuality of nature. For most scientists, the difference is not important, but it is one that has filled the philosophical journals for 50 years.

that one thing causes another. For what exactly is the relation of cause and effect? We may see that B always follows A, and may also find that if there is no A, then there is no B. According to the celebrated philosopher David Hume (1711–76) causation is nothing more than the human mind noticing a regularity, and jumping to the conclusion that something more is involved, namely, a mechanism. We should keep in mind Hume's scepticism in the following discussion, which attempts to show that biological explanation is very dependent on causation, but, before that, I will examine an actual biological explanation. My example is taken from protein synthesis:

> Information held by a strip of DNA is transcribed as messenger ribonucleic acid (mRNA), which then moves physically out of the nucleus and into the cytoplasm. Here, the mRNA is brought into intimate contact with ribosomes, transfer RNA (tRNA) and amino acids. As a result of this contact, the information carried by the RNA is translated into strings of amino acids, which subsequently fold up to form proteins. Later, at diverse places in the cytoplasm (for example the Golgi body) the proteins are given additional treatment, prior to their being secreted from the cell.

Clearly this is a reasonable, if basic, explanation of protein synthesis, but it is not final. For example though it is known that chains of amino acids fold up to form proteins with a three-dimensional structure, little is known about how this is done, or how it is controlled. This is work for the field of proteomics. Research here will generate additional, myriad details of molecules moving from place to place, bonding and uncoupling, but the richness of proteomics will fit into the basic chain of causes and effects described above. The explanation, therefore, has a co-ordinating role: it provides a satisfying and useful account of certain events, but also suggests further research.

You can see also that my cellular explanation does not mention any laws, but that there is a long list of events, apparently connected up into some kind of a story. It looks likely that biological explanation is going to be closer to Scriven's identification of explanation with causation, than to Hempel's advocacy of the use of law. Even at the level of the DNA and its pattern of nucleotides being 'transcribed' into mRNA, the laws of physics and chemistry cannot be used to predict how a strip of DNA will be used. It is worth remembering that when the genetic code was worked out, it was by experimentation with microbes, not by using the laws of bonding.

In conclusion, we can see that biological explanation relies on causation. When we read about biology, we soon come across cause and effect. A gives B, and B gives C, and C gives D. To explain D, you show how A, B and C led from one to another, causing each other, finally giving D. We see these chains in the citric acid cycle, or in the oxygenation of haemoglobin as blood passes through the lungs, or in those enzyme metabolic pathways where a product becomes a substrate for the next reaction. Such explanations are highly characteristic of molecular biology, and because molecular biology dominates biology, for the student faced with an examination, the main task is usually dominated by learning the links of these causal chains.

2.6 Metaphysics and biology

There is a sub-section of philosophy called metaphysics. This is the field of inquiry that investigates the structures upon which knowledge, including science, can rest. To a philosopher, it is interesting that causation plays such a large part in biological explanation. What is interesting is that the concept of causation does not seem to be a very robust foundation on which to build scientific explanation. Or to put it another way, if causation is a major part of your explanation, then you should be cautious in claiming that you are describing reality. Metaphysics does not question the usefulness of the concept of causation; it simply notes that it is hard to see what causation is, other than the habit humans have of making generalisations.

Metaphysics need not be a problem for biology, however much we use causation in our explanations. For philosophers now take for granted that explanations are not only incomplete, but can be reliable even if they cannot be grounded in one universal, foundational reality. However, I now want to change the direction of my argument, and put the tricky issue of causation to one side. For it is not true that biological explanation is only a matter of cause and effect. It is characteristic of biologists that, very often, they rely neither on causal explanations nor on laws. Instead, explanations for a phenomenon often involve cobbling together strands of research from different contexts. The following example shows what I mean. When we investigate a blood protein, the explanation for its presence cannot properly be considered simply a matter of chemical reactions, however complex. In a real sense, the explanation for the presence of that blood protein must include the story of its own evolutionary history, and its contribution to the adaptiveness of the organism. That account may well be a well-accepted part of biological knowledge, but bearing in mind the lengths of time involved, the lack of a fossil record, and the difficulty of re-running evolution, the evolutionary account of the blood protein is more likely to resemble prose in a history book than a diagram of a chemical metabolic pathway. In other words, we partly account for ('explain') a biological trait like a type of behaviour by considering its function in the animal. If we see that a brightly coloured cuticle makes an insect very much less likely to be mauled by a bird, then we consider the explanation for that coloration rests in the theory of natural selection. It would seem oddly unbiological to explain the cuticle only in terms of the chemical processes that made the coloured pigment.

Biologists, it seems, work simultaneously at different levels. When they want to explain a structure, even a molecular structure, they refer to a number of different levels. We can consider these levels to be the molecular, the cellular, the physiological, the organismal and the populational. Each is important; each depends on each other. There is an extra dimension too: evolution. All biological entities have changed over time, some more than others. What we see around us now is, to a lesser or greater extent, the result of natural selection. So it is quite usual to find explanations in biology drawing from different levels, and from evolutionary theory. For example a molecular biologist explaining a metabolic pathway might investigate the substrate–product–substrate relations between a series of molecules, then refer to the uses an organ makes of the secreted material, consider how a deficiency in the chain can disable an organism, refer to the metabolic chain in different cells or in different organisms, and also make some evolutionary comparisons. This is arguing across levels: it is building a rounded, biological explanation. The reductionist perspective is needed, but so is the perspective of the biologist interested in physiology, or whole organisms, or the relation between organisms and their environment, or their evolution. It may well be that a satisfying biological explanation is not one that invokes laws, nor one that simply lists a few causes, but is one that uses different levels, and so unifies – or makes connections between – some of the varying research styles.

Committed reductionists will consider that the most important of these levels is the molecular. They will argue that while for the moment we might reluctantly include references to the higher levels, in time this will be unnecessary, and an explanation at the chemical level, perhaps using a new set of biomolecular laws, will suffice. Anti-reductionists disagree. They do not dispute the revelatory powers of molecular techniques, but they argue that the higher levels cannot be ignored.[6] Understanding of these higher levels, the argument goes, cannot be replaced by – reduced to – theories that apply to lower levels. The commonest justification of this position is simply to point out that higher level phenomena often cannot be predicted from a knowledge of the component parts that constitute the lower level. Take water for example: its behaviour cannot be predicted simply from a knowledge of hydrogen and oxgyen. For the anti-reductionist, this inability is not simply a result of ignorance. It is a matter of principle that every level has its own characteristic patterns and processes, particular to that level, and not replaceable by the patterns and processes of a lower level. In conclusion, it can be seen why the issue of reductionism links up to the question of explanation. The more overtly reductionist the biology, the more overt will be the reliance on causal explanations, and causality, philosophers have found, is not a very sure foundation for truth. Yet when the causal accounts are complemented by an anti-reductionist perspective, the quality of the explanation is enriched and stabilised: by drawing on different branches of biology, the explanation becomes more reliable.

2.7 A review of terminology

If you read further into the philosophical literature covering explanation and reductionism you are likely to encounter a number of technical terms and beliefs. What follows is a quick guide.

Monism

Earlier we briefly described the beliefs of the philosophical reductionist. We can now use another label to describe this way of looking at the world: monism. Monism is the belief that the world is made of only one kind of thing, that however complex and disunited the world seems, there is a basic uniformity. For a contemporary monist, physical particles are the most important – the 'most real' – aspect of the material world. A monist believes, therefore, that science should aim to understand the world in terms

---

[6] In his book *The Disorder of Things* (Cambridge, Mass.: Harvard University Press, 1993) John Dupré uses metaphysics to challenge reductionism. He argues that, contrary to the philosophical reductionist view, cats and dogs are as fundamentally real as atoms. The study of organisms, according to this metaphysical position, should, therefore, not be under pressure to find genetic or chemical explanations. On the whole, biologists do feel that pressure: their jobs may depend on it. Perhaps if some metaphysicians were appointed to the funding councils, things would change.

of these particles and the investigation of these particles; though cells, organs, organisms, populations or societies seem like a different kind of thing from an atom, this is an illusion. Monism, therefore, urges that we adopt a reductionist research programme, because this is the only route to a proper understanding of reality.

Dualism

Like the monist, the dualist sees the analysis of physical particles as the only reliable basis for knowledge, but the dualist makes an exception for the understanding of mind (and, therefore, consciousness). The mind is separate from the body; it has its own rules, indeed, its own level of reality. We have all sorts of names for that level: soul, spirit, mind and consciousness. Dualism as a philosophy is usually identified with the French philosopher René Descartes (1596–1650) and is popularly known as the 'mind–body split' or the 'mind–body problem'. That there is something misconceived about dualism is universally acknowledged: after all, the mind affects the body and vice versa. We see this every time we blush at some embarrassing memory. The problem here is that if mind and body are separate branches of reality, how can a memory cause a blush? I mentioned earlier that philosophers, in considering the nature of explanation, find many problems lurking in the apparently simple idea that 'A causes B'. The root difficulty is in determining the nature of the link implied by the word 'cause'. In the case of the dualist, the problem is particularly acute. The mind makes all sorts of decisions, and causes appropriate movements of the body; but if the mind is believed to be a fundamentally different sort of thing from the body, then it is understandable why philosophers find it hard to see what the mechanism of causation could be. (The concept of dualism is discussed further in Section 4.4, in the context of animal rights and ethics, under the heading 'Animals are simply machines'.)

Pluralism

The pluralist differs sharply from the monist and the dualist. In the pluralist's world view, the world is not best understood as consisting fundamentally of one kind of thing (physical particles) or even two (physical particles and mental states): instead, there are many. In biology, it is normal to find scientists devoting their attention to particular levels. The pluralist is happy to give equal status to various levels of investigation, for pluralists see reality as made up of many co-existing levels, each as real, and as fundamental, as the other, and including the cell, the tissue, the organism, the population. The philosophical reductionist, who we have now redefined as a monist, urges that we drive our inquiries towards the lowest level (the particle), but the pluralist dismisses the implication that cats, trees and beating hearts are entities less real than atomic particles. Nor is the classification made by dualists acceptable, for in that philosophy, the world is split into two basic kinds: the material and the mental.

Organismal biology

At the start of the chapter, I suggested that for many biologists, it was organisms, and not molecules, that first got them hooked on the subject. Organismal biology is the home for those researchers who still investigate the lives of organisms. They are interested in whale song, coloration in plants, the burrowing behaviour of earthworms. The characteristic of organismal biology is a determination to understand the lives of whole organisms, in all their variety. Clearly, part of that understanding will come from molecular and genetic studies, but 'whole-organism' biologists share a common philosophy, that reductionism cannot fully describe the life of a growing, respiring living thing. For example a physiologist studying the effects of adrenalin on mammalian organs can measure the precise way in which adrenalin influences sugar uptake, vasodilation and heart rate. Yet all this information, however ingeniously obtained, and however reliable, loses some significance unless it is placed in the context of the life of the organism.

Emergence

Organismal biologists simply like animals and plants. Even though from time to time these biologists mix reagents and measure events at the level of the molecule, they are always glad to return to their work with actual organisms. In their preference for working at the level of the organism, they might gain support from another philosophy, sometimes called 'emergentism' or simply 'emergence'. This is the belief that each biological level has properties that can neither be predicted, nor explained, by reference to a lower level. Thus a whole animal has properties that depend not only on components, but on the way these components are organised, and 'collaborate'. The behaviour of an organism cannot be predicted, or fully explained, by a knowledge of organs, cells or genes (though a knowledge of lower levels will make a contribution). The word 'emergence' is not meant to be mystical, as if some strange force or field applies to living things, but not to components; it is simply meant to imply that new properties emerge as you move from level to level, and that an attempt to explain these properties in terms of lower levels will end in failure. A biologist interested in the concept of emergence is likely to entertain quite strongly anti-reductionist ideas, believing perhaps that in time we will discover biological laws and principles that apply only to these higher levels, and cannot be linked back to (for example) genetics. This makes modern biology, with its emphasis on genetic causes, a slightly unwelcome place for those investigating emergent properties. Yet emergentism is not a New Age philosophy, involving spirits or the supernatural. There is no unmeasurable or unknowable soul that animates the upper levels, but leaves the atoms and the molecules alone. For the biologist interested in emergence, it is simply that highly complex organisations (as found in biological systems) generate phenomena in ways that cannot be understood simply by measuring the components. In time, according to the biologist interested in emergence, complex laws referring to the higher levels will be discovered, thus making reference to atoms, molecules and genes far less of a priority.

Vitalism

Vitalism is now a dead philosophy. Its period of maximum influence was at the start of the twentieth century. Vitalism was certainly anti-reductionist, but not simply because it takes the higher levels seriously; for unlike the ideas of emergence, vitalism sees special forces at play in living things. For the vitalist, living things are possessed (literally) by a 'life force', utterly distinct from the physico-chemical forces so far discovered. The biologist who we associate most with vitalism is Hans Driesch (1867–1941). He was a superb embryologist, extremely dextrous in his anatomical investigations of embryos. Historians suggest that it was his biological skills that led him to suggest the existence of a life force. At the time, it was customary to look for chemical and even electrical causes for biological phenomena, but looking at the emerging, swirling shapes of his developing chicken embryos, Driesch could not see how electricity could explain the amazing creation of organisation. However, his alternative explanation, vitalism, suffered from being impossible to investigate: how do you investigate a life force? Driesch abandoned biology for philosophy but lived to see biochemical methods begin to be applied more successfully to his field.[7]

Holism

Our final philosophical tag is holism. Perhaps this is best defined by the motto 'The whole is greater than the sum of the parts.' Holism is suspicious of reductionism; it prefers to study whole systems, to emphasise connectivity and co-operation. It is not a popular term in academic circles. Scientists and philosophers swimming against the tide of modern biology are happier with the tags mentioned earlier: emergence, pluralism and anti-reductionism. Perhaps the problem is caused by the association of holism with allegedly anti-science movements, such as holistic medicine.

2.8 Anti-reductionist biology today

So far, this section has summarised a few philosophical positions relevant to our discussion of reductionism. It is worth considering too whether, in modern biology, there are research programmes that are either determinedly anti-reductionist, or try to maintain an interest in higher level phenomena. We will consider, again very briefly, two of these: the Gaia Hypothesis and the Complexity Theory.

The Gaia Hypothesis

This hypothesis was articulated in 1974 by the atmospheric scientist James Lovelock and the microbiologist Lynn Margulis (whose work on symbiosis I have discussed in Chapter 1). Both scientists were (and are) at home with the minute components of organic life. Lovelock invented machines for detecting minute levels of gases in atmospheres, while Margulis became well known for her ideas on the evolution of cellular organelles (such as mitochondria and chloroplasts). Thus, both are interested in atoms and molecules, not simply in roaming mammals. Yet Lovelock's understanding of atmospheric composition led to an idea that is far from reductionist: that the physical and the biological realms are unified in a system of

---

[7] Vitalism may be dead, but its significance in the history of biology has been dealt with sympathetically by the evolutionist and historian Ernst Mayr. His book *The Growth of Biological Thought: Diversity, Evolution and Inheritance* (Cambridge, Mass.: Harvard University Press, 1982) puts vitalism into its context, and shows why the history of biology is an important intellectual resource for working biologists. Discussing the role of vitalism in the early nineteenth century, Mayr remarks (p. 847) 'It is curious how often erroneous theories have had a beneficial effect for particular branches of science.'

feedback loops. According to the Gaia hypothesis, life on earth adjusts the state of the physical environment, so as to make it more life-friendly.

It might seem a trivial idea to suggest that organisms affect their physical environment. What else is a bird's nest, or an earthworm's burrow? The Gaia theory suggests something much grander: that the global composition of the earth's atmosphere and its temperature are controlled homeostatically through the action of organisms. Most organisms shift their environment far from equilibrium. It was Lovelock's idea, when contracted by NASA (the US National Aeronautics and Space Administration) to investigate ways of locating life on alien planets, simply to study the atmosphere. Disequilibrium could be taken as a sign of life. In the earth's case, the atmosphere is in an extreme state of disequilibrium, with levels of oxygen and methane being far in excess of what would be expected. The Gaian idea of homeostasis points out that, at 21%, the level of oxygen is just below the fraction at which fires would disrupt land life. It is also of interest to Gaia theorists that the solar energy arriving at the earth has increased by 25% since life began 3.8 billion years ago, but there has been no corresponding temperature rise in the environment. For that to happen, more energy is being reflected back into space, or less is being absorbed (for example by the carbon dioxide fraction of the atmosphere). The Gaia Hypothesis suggests that the regulation of the atmosphere is being achieved 'by and for' the biosphere. Those scientists funded to study global self-regulation (and there are rather few) search for feedback pathways between organisms (especially microbes) and the physical environment.

Clearly, Gaia is not a purely reductionist research strategy. It is working at a very high level indeed: a level that conjoins the global environment and microbial populations, for example. Sometimes, even, Gaia is described as declaring the earth to be a 'superorganism', and there have been criticisms by evolutionists that the Gaian biological entities (such as feedback loops between oceanic microbes and the atmosphere) are inexplicable in terms of Darwinism. What is clear is that scientists investigating Gaia have to be competent at physics, chemistry and biology. In other words, they have reversed the basic trend of research science, which is towards specialisation.

Complexity theory

This is our second example of a research programme rejecting a fully reductionist strategy. The ideas stem from one simple yet profound observation about biology: that living things are extraordinarily ordered. In the eighteenth and nineteenth centuries, it was thought that the study of biological order, or form, revealed not only the fabulous intricacy of living things, but also something of the powers of God. An essential skill was to be able to describe and draw the shapes and patterns of living things. Few biologists today are good draughtsmen; very likely even fewer consider the shapes of nature a direct example of the power of a deity. The basic assumption is that Darwinian evolution explains the organisation we see in organisms, whether it be the layout of microtubules in a flagellum, the whorls of a moss, or the limb pattern of the tetrapod. In other words, the shapes we see are adaptations modified over time in a line of ancestors. Furthermore, the generation of these forms in the actual individual (sometimes called ontogeny) is largely a matter of the right genes being expressed at the right time. This idea of biological form is partly reductionist because it looks for explanations in genetic pathways, but a higher-level explanation is also involved, one involving Darwinism, and thus ecological interaction.

There is plenty of order, and transformation, outside biology. The snowflake is an example, yet its order emerges from the vibrating water atoms within a rain drop. From that disorderly activity, a crystal of very definite form emerges. Clearly, laws about atomic bonding are involved, and other constraints must be important too, for example gravity. Thoughts such as these about snowflakes are now also being applied to biological systems. Theorists working in this area accept that form has evolved over time, and that genes are involved in the development (and of course the evolution) of organisms, but they see in the patterns of life signs of simplicity emerging from complexity, or even from chaos. When a plant shoot develops a spiralling pattern of leaves, could the order be being self-generated, as with the snowflake springing from the water drop? The idea is not simply that symmetry and simplicity can emerge from chaos, and organisation from disorder. The claim is that there are a series of higher-level (but unknown) laws that make life inherently self-organising. This is a controversial, anti-reductionist idea, downplaying (but not rejecting) the theories of Darwin and molecular genetics.

I have implied in several places in this chapter that reductionist biology tends to be very specialist and narrow. For example my account of the Gaia hypothesis imagined a troop of anti-reductionists skilled in a variety of scientific fields. For many decades scientists have accepted that, to have any chance of doing original research, they must refine down their expertise, so that perhaps only ten other people in the world can understand their research papers. Recently though, there has been a counteracting movement. New ideas often flourish at the boundaries between the disciplines: between astronomy and biology, or physics and biology, or computer science and biology. Nurturing these ideas is hard; they often have no home department, perhaps no journal. These so-called interdisciplinary initiatives are often, however, very fruitful. They redefine research areas, and attack subject boundaries. They are prepared to take up anti-reductionist lines of inquiry, and to establish their own, new levels of explanation. For a trainee scientist, these new interdisciplinary scientists are worth watching.

2.9 A case study: phenylketonuria (PKU)

There are examples of genetic diseases where the links between gene and symptom are well understood. Phenylketonuria (PKU) is one such disease. The story of PKU is a self-evident tribute to reductionist medicine, showing how one of the most common inheritable metabolic diseases, which is a severe affliction, can now be avoided if detected soon after birth. PKU patients used to make up about 1% of the severely mentally retarded patients living in institutions. Babies born with the condition showed few symptoms but within a few months began to suffer mental deterioration. Thereafter, PKU patients rarely reached a mental age greater than 2 years. Motor co-ordination was affected and patients rarely talked. Other symptoms included irritability, violence and fearfulness. Before the chemical roots of the problem were identified, patients rarely lived beyond 30 years.

PKU is caused by a defective gene. The mutation is recessive and autosomal (i.e. it is not on a sex chromosome). The disease is found in around one in 10000 babies born in Europe and the USA. It is more common amongst the Irish and the Scandinavians, but is much rarer in Asia. The effect of the mutation is to disrupt the production of the metabolic enzyme phenylalanine hydroxylase (PAH). The usual site of action of this enzyme is the liver, where it converts the amino acid phenylalanine (part of a normal protein-containing diet and an essential amino acid) to tyrosine. Without the enzyme, phenylalanine builds up towards a high level, and is transaminated to phenylpyruvic acid. It is the excess of this acid, rather than the lack of tyrosine, which causes the mental deterioration.

Treatment for PKU arrived in the 1950s when doctors began to treat patients with a highly restrictive diet, one that contained no high-protein foods, and therefore, a low content of phenylalanine. It is an unpalatable, unattractive diet, but as a treatment it is extremely effective. What is important, of course, is that treatment begins very early in life, before the degeneration sets in. Yet the diagnosis of PKU was done by observing the symptoms, and these were caused by irreversible brain damage. What was needed was a test that could be done early in life. For a baby with PKU, treatment that starts at 6 months of age is already too late. Even at 3 months, there may be little benefit, with brain damage already extensive. Fortunately, the development of a cheap test for blood phenylpyruvic acid in 1963 (the Guthrie test) allowed the possibility of screening at birth, and this is now one of the routine tests carried out on newborn babies.

This looks like a self-evident example of the benefits of a reductionist approach to a medical problem. Some of the links between gene, enzyme and cell chemistry have been determined, the classical genetics of the disease in the population is fairly well known, and the gene is known to be carried by 1 in 50 people, which is a rate similar to that of diabetes. Today, we not only have the classical genetic explanations of the disease, we also have both the metabolic explanations and some understanding of the gene involved, which has been produced using the tools of molecular biology. For example the PAH gene has now been mapped to the end of the long arm of chromosome 12; it has about 90 000 base pairs, which makes it a large gene, and it codes for a protein of 451 amino acids.

However, as we know, science is never complete. The example of PKU shows this well. On the one hand, my superficial account suggests that PKU is a problem almost solved, thanks to the establishment of the cellular chemistry at fault and the gene concerned using the techniques of molecular medicine. On the other hand, a slight change of viewpoint reveals how little is known about PKU. For example it is still not clear how an excess of phenylalanine can produce those fearsome symptoms of brain damage described earlier. A further interesting question arises from the fact that there are a number of types of PKU, with symptoms of differing severity: why? Certainly, at the level of the gene, there are a number of nucleotide polymorphisms, which indicates perhaps that there are a number of mutations linked to PKU. Yet there is no correlation between the particular polymorphism a PKU patient carries within his PAH gene, and the type of PKU suffered. It is known that the metabolism of phenylalanine involves several enzymes and coenzymes; indeed, PKU is a problem of varying severity, where the degree of impairment of phenylalanine conversion varies rather markedly from one patient to another. It can be argued that this is simply 'a gap in the knowledge' that will be filled. Perhaps, but at present we have to admit that PKU shows how reductionism offers powerful, but incomplete explanations. An example of how these limits have social, as well as scientific, dimensions is shown by a problem that arises solely from the successes in explaining and treating the disease: it is known as the problem of maternal PKU.

Only recently has it become common for a woman testing positive for PKU to have a baby. Previously, such a woman would have been severely retarded; probably she would have been living in an institution. PKU patients did sometimes have children, and because of the high phenylalanine levels in their blood, the fetal blood was also affected, resulting in brain damage. The problem that sometimes occurs now, in these days of testing and screening, is that a woman may not know that she has PKU. This apparently unlikely scenario is possible because, in the last 30 years, babies tested positive were given the special diet, and this continued into childhood. Then, perhaps at the age of 6 years, the treatment would be stopped. Serum phenylalanine returned to a high level, but with little or no effect, presumably because brain development was now established. No doubt this suspension of treatment would have something to do with the diet being so unappetising; and indeed it seems as though some such children, when they had grown up, had no memory of any diet. Furthermore, they might not even have been told of their condition. An article in the Boston Globe put it like this: 'Some families put off vacations and restaurant meals until the child was off the diet. Nothing pleased them more than the feeling they could forget the whole thing.'[8] Even though they might indeed forget the whole thing, the high phenylalaline levels return, and in the event of pregnancy, the phenylalaline crosses the placenta and interferes drastically with the developing fetal brain. An ethical question arises, therefore: should women with PKU be monitored through life, and required to stay on the diet? – current advice is that coming off the diet after brain development has finished has some minor side effects. The fact that there is an ethical issue arises partly because the simple reductionist explanations we might perhaps expect are not available. For example though maternal phenylalanine certainly crosses the placenta, much remains unknown about the actual relationship. The question has been asked: could a very low level of phenylalanine, as would be caused by

[8] Caldwell (1981) quoted in *Genetics: Human Aspects*, 2nd edn, by Arthur P. Mange and Elaine Johansen Mange (Sunderland, Mass.: Sinauer Associates, 1990), p. 311.

over-treatment of the mother, itself damage the health of the fetus? We can see that in PKU, we have an example of a medical condition whose proper management, both for the individual and for society, is not merely a matter of the proper knowledge of molecules. It is also a matter of sensitivity and insight.

2.10 In conclusion: the problem of determinism

One philosophical concept remains to be discussed, that of determinism. Put at its strongest, this is the belief that the causes of all future events can be found in the events of today: that 'What happens must happen'. Popularly, determinism pictures the universe as a giant cosmic mechanism, every component geared and connected. Constructed and set in motion when the world began, the precise future of each component is fixed. The opposite of this extreme philosophical position would be the universe where events just happen, without any causal connections to events happening earlier.

The truthfulness (or otherwise) of determinism, and its usefulness in telling us how to do science, is a problem for philosophers. The strong version of determinism, as suggested by the image of the cosmic clock, is not taken very seriously. There is more interest in incorporating the ideas of probabilities into laws, explanations and causes. For example a law of nature could be viewed as a statistical regularity. Certainly, for explaining the events of ordinary life, statistics seem more useful than fatalism. A good example is smoking. No one doubts that smoking causes cancer, but equally, we are familiar with the idea of someone smoking every day of their adult life before passing away peacefully at the age of 87.

The link between reductionism and determinism is probably quite clear to you. If you are to know the motion of the clock, then to start with, you must identify and measure the components. With the components understood, the motion of the mechanism will be understood. This may be impossible to achieve in practice (or even in principle) but it is a seductive idea. In biology, the idea can be seen in the many headlines and press statements describing genes for Alzheimer's disease, breast cancer, homosexuality, depression, musicality, and fitness at high altitudes. The news stories about genetics generally focus on genes that have been linked to a disease. More controversial are the sightings of the 'gay gene' or the 'intelligence gene'. Here, we not only see reductionism with complex sets of behaviour being defined as traits analogous to eye colour, or types of skin, which are explicable by referring to a strip of DNA; we also see determinism – to have the gene is to have the ailment (or IQ level, sexual orientation, perhaps even political leaning). The more a biologist believes that possession of the gene inevitably leads to the condition, the more biologically determinist that biologist is; and the greater the belief that control is caused by the environment, the more environmentally determinist the biologist is. Not surprisingly, given the politics implied, the status of biological determinism is much more controversial than its more subtle cousin, reductionism. For if we are very strongly determined by our genes, then education programmes, drug rehabilitation centres and even peace movements, are either a waste of time or must be considered difficult, because they contradict human nature (assuming of course what seems most unlikely, that there are genes for IQ, alcoholism, and aggression).

Those simple causal explanations so favoured by the molecular biologist do indeed open up a whole world of cellular knowledge, with all the associated possibilities for the pharmacologist. There are numerous examples where events involving DNA, RNA, ribosomes, polypeptides and enzymes have been linked together. We have a model of the chemistry of the cell, and it is constantly being elaborated. Yet as we saw in the case of PKU, it is hard to build out from this model to the level of the whole organism, to its behaviour, appearance and diseases, and to the feelings of the patient. The problem is that causal explanations, which are the justification of determinist beliefs, are at their most compelling in biology when they look at molecular events; but biological molecules have evolved, and the kind of explanation used in evolution is not a simple causal explanation. In evolution, it is the existence of particular molecules that first catches the attention: they have arisen and persisted by a combination of mutation (a molecular event) and selection (an ecological, long-term event). The first is quick, the second slow.

A key reductionist idea is that of the genetic programme. It is also the basis of a major determinist theme in biology. You will know of accounts where genes are described as the blueprint of life. They are the secret of life; they are said to control the cell. According to one version of Darwinism, organisms are simply vehicles for passing on genes. The metaphor encourages us to see in genes the primary motor of life, and with the results of the Human Genome Project, genes are being linked to aspects more complex than a particular enzyme in the digestive system. It is intelligence, sexuality and future health that are being reduced to the genetic level. Yet we have begun to see in this chapter that caution is appropriate: genes are not so easily defined – they have a role, but so do a range of influences working at a higher level. Genes are not simple entities, strips of DNA. They seem to be more complicated than that. Alternative splicing, where different zones of a section of DNA can be transcribed into mRNA to produce different peptides, suggests that one gene can have more than one function. We know too that genes can be assembled from bits of DNA scattered across different chromosomes. Many genes may be involved in one trait, as seems to be the case with the so-called 'polygenic diseases', such as heart disease, and one gene may be involved in many traits. To muddy things further, we have to accept that the machinery involved in controlling genes is very complicated, and is certainly influenced by events in the cytoplasm. Also, as regards the development of the individual, the forms that take shape may be shaped not simply by the genetic programme, but by influences working at a higher level of complexity. In the debate about genetic determinism no one disputes the importance of genes: what is being contested is how knowledge of genes, and knowledge of society, relate to one another.

3

Evolution

3.1 A philosophical introduction

In the last chapter we saw that reductionism, for all its successes, has disadvantages. Its benefit is that, by focusing always on the minute components of biological processes, reductionism produces those causal explanations so valuable to research biologists in general, and to medical researchers in particular. The cost comes when the larger picture is undervalued, or even ignored; when the whole, living, organism is forgotten in the drive to catalogue its parts. I said, therefore, that hard though it may be to escape the clean confines of reductionist biology, the corrective comes by taking a wider perspective – by lifting your eyes from the test tube, and looking around. You do not need to look far, for in the life sciences, the largest picture of all is provided by evolutionary theory. So, having spent the last chapter appreciating reductionism, but also cautioning against the narrowness it can foster, I will now look at the characteristics of evolutionary theory that put it at the heart of biology.

Evolutionary theory is a loose collection of ideas surrounding the idea of change in living things. It is preferred territory for an amazing – and amazingly disunited – company. Indeed, it is not hard to find evolutionary ideas pressed into supporting contrary arguments: that people are naturally aggressive (or peaceful); that the pace of evolution is even (or irregular). I shall cover some of the disputes of evolutionary theory and in the process, I hope, portray its enduring achievements. It is an area where reductionism and anti-reductionism work together profitably. For though the evolution of life comes about through the most complex of relations, and is life considered on a grand scale, reductionist biology provides many vital insights into its processes. One small example will suffice. In Darwin's theory of natural selection, genetic variation amongst the members of a population is a fundamental necessity if evolution is to take place. Darwin did not know what caused the variation to be generated; though this was an inconvenience to him, he was comforted by the thought that soon enough (perhaps after his own death) the source of variation would be known. Indeed, biologists do now know, and ascribe variation to mutation, crossing over and linkage. That knowledge is reductionist, relying on an understanding of chromosomes and biochemistry. Moreover, now that genomes can be sequenced, 'gene trees' are working alongside the more familiar cladograms produced by the analysis of morphology. So evolutionary theory should not be seen as the triumph of non-reductionism. Rather, it offers modern biology an example of how the two great traditions can work together. There may be some conflict, but as I shall show, it is creative conflict.

Evolution is the word we use to describe something fundamental about life on earth: that it has changed over time. Today, evolutionary thinking is still dominated by the work of Charles Darwin, who died in 1882. Darwin's theory, in its modern form (neo-Darwinism), is the favoured account for explaining how evolution takes place. It is easy to forget, therefore, that evolution, and Darwinism, are two separate things. Evolution is simply the fact that life has changed over time; Darwinism – the theory of natural selection – is the main explanation and so forms the central theme of this chapter. In exploring that theme, I will describe some of the controversies that, from the beginning, have surrounded Darwinism. When thinking about these issues you may find it useful to remind yourself that the fact of evolution, and its explanation by the theory of natural selection, are not the same thing. Keeping the two in a harmonious relationship has been the job of Darwinians for the last 150 years.

This chapter will first ask some general questions about evolution. We need to appreciate something of the time scales involved in the evolution of life, and to feel how crucial an insight it was when scientists became able to track back in time to the formation of the earth. It is clearly important that the fossil record – a fixed, dead memory – can be interpreted as revealing the living, evolving nature of biological populations. Like history itself, fossils tell us about the present, not just the past. Our discussion of evolution will give an impression of the field's vast sweep. Evolutionists draw on a range of evidence, from the genetics of living creatures, to the distribution in rocks of fossils formed 500 million years ago. Yet we will find ourselves returning to the same themes again and again. How can we know about changes that perhaps took place hundreds of millions of years ago, events that cannot be repeated in laboratories, and that in some sense cannot be observed? Do all these changes, to which we give the term evolution, and which led to the development of complex nervous systems and more latterly *Homo sapiens*, allow us to deduce that the story of life on earth is one of progress? The chapter, finally, will focus down on Darwinism itself, examining some of the philosophical questions surrounding Darwin's great idea, the theory of natural selection. To what extent can natural selection explain the wondrous diversity and finely tuned adaptation we see around us? Can Darwinism, with its famous motto about the 'survival of the fittest', explain altruism as well as self-centredness? Lastly, perhaps most controversially of all, can Darwinism make a worthwhile contribution to the way that we understand our own behaviour, whether selfless, criminal, loving or insane?

3.2 The work of Charles Darwin

Charles Darwin was born in 1809, the son of a wealthy doctor. Our familiar image of Darwin is as a very wise old man; in the best-known photographs he resembles an Old Testament prophet, like Moses. Yet Darwin was never a conventional scholar; school he considered of very little use. In his autobiography he wrote: 'Nothing could have been worse for the development of my mind than Dr. Butler's school as it was strictly classical, nothing else being taught, except a little ancient geography and history. The school as a means of education to me was simply a blank.'[1] University was not much better. He dropped out of medical training at Edinburgh, and muddled his way through a theology course at Cambridge. In the end it did not matter: Darwin never had a job in his life because he had no need, for he had a good private income. He also had the kind of interests, and the temperament and health that do not fit well with employers.

As a naturalist Darwin was self-taught. Both at Edinburgh and Cambridge, he was assiduous in turning up at public lecture series given by eminent naturalists. He nurtured friendships with geologists, botanists and ornithologists. He went on holiday to Wales to explore the theories of geology and found his own researches more rewarding than anything the Edinburgh and Cambridge academics could offer him officially. The natural result of this was that Darwin learned a lot of science

[1] *The Life and Letters of Charles Darwin, including an Autobiographical Chapter*, 3 volumes, edited by F. Darwin (London: Murray, 1887).

very quickly. He became known as an excellent student of nature, even if a worry to his father. In 1831, at the age of 23, on holiday at home enjoying the hunting season, Darwin got an invitation to travel round the world in an Admiralty ship, HMS *Beagle*, to keep the captain company but with the opportunity to do as much natural history as he liked. Darwin liked the idea, a cautious father was talked round, and in December 1831 Charles Darwin set out on a trip that would change his life, and the subject of biology, forever.

This trip was Darwin's *wanderjahre*, his youthful wandering. He travelled round the world, seeing exotic cities and peoples, climbing the Andes, falling ill, and writing a book. He collected wherever he went. Crates of rocks, fossils, eviscerated corpses and dried insects accumulated in his 6 foot by 6 foot cabin. Every 3 or 4 months, the crates would be loaded on to a passing ship bound for London and sent back to be pored over by Darwin's impressed and envious friends. Darwin filled notebook after notebook, starting a habit of recording and writing that he kept up till his death, a habit that today absorbs the full effort of the several scholarly teams who labour to transcribe and edit the mountain of Darwin's correspondence.

Darwin returned to England in 1836, after 5 years at sea. He had undertaken an extraordinary journey, visiting four continents and travelling 20000 miles. The record of that journey is Darwin's *Journal of Researches*, one of literature's most vivid and authentic descriptions of the joys of being young, a traveller, and keen to experience as much as possible.[2] In spite of his later reputation as a father of science, the young Darwin was a man of sensibility, outgoing rather than introverted, and he relished travel. On his return, however, he was to lead a more and more reclusive life. It is extraordinary to think that although he was a man who travelled more than most modern people, who had experienced typhoons, near shipwreck, earthquake and volcano, who had seen slavery and murder, the thrashing of drunken sailors, and the systematic torture and slaughter of south American natives by the same *gauchos* to whom he owed the security for his gallops across the pampas, and who had endured day-long bouts of seasickness, 3-month intervals between letters, and a 5-year separation from the family he loved, that not once after the age of 24 did he cross the English channel.

[2] This was originally published in 1839 as *Researches into the Geology and Natural History of the Various Countries Visited by H.M.S. Beagle* (London: Henry Colburn Publishers), and has since been published under various titles, including *Voyage of the Beagle: Charles Darwin's Journal of Researches*, edited by Janet Browne and Michael Neve (Harmondsworth: Penguin, 1989).

You can visit Darwin's safe haven now, Down House in Kent, and walk the Sandwalk, a chalky path he paced each day, a man of complete routine. He had replaced his outer, diverse, wildly various life with one of complete harmony and apparent tranquillity, periodically punctured by a mysterious illness that lasted all his life. Yet if you read his letters you will see an emotional man, urgently exploring a big idea, not afraid to judge his fellow professionals against his own views.[3] He had decided, by the end of his travels, that life had evolved, and that species were not fixed, but changed over time. There were others who had said this, and had even published theories for the evolutionary mechanism. Yet these theories were more-or-less supernatural: God was involved, or an 'inner drive for perfection'. Darwin wanted a theory that relied on observable, natural phenomena, not on idealistic and romantic conceptions of nature. He wanted a mechanism, not a mystical belief; he would search for a mechanism that could be written in ten lines, and be understood by a 10 year old.

Darwin's critical experience was that youthful trip around the world. Though he was to spend 25 years reading and reflecting on evolution, pursuing a correspondence that spanned the world and that furnished him with details of organisms in five continents, the basis of all this work was the voyage of the *Beagle*. After that journey, he wrote to the captain of the ship, Fitzroy:

> However others may look back to the Beagles voyage, now that the small disagreeable parts are all well nigh forgotten, I think it far *the most fortunate circumstance in my life* that the change afforded by your offer of taking a naturalist fell on me – I often have the most vivid and delightful pictures of what I saw on board pass before my eyes. These recollections and what I learnt in Natural History I would not exchange for twice ten thousand a year.[4]

Darwin kept his kernel of an idea to himself for two decades. During that time, he slowly worked on it, shaping it. Darwin used his correspondence,

---

[3] There are more than 14 000 letters, collected in a dozen or so volumes. Darwin may have been reclusive, but he understood the need to gather information and to debate ideas. Unwilling to walk further than the end of his garden, Darwin nonetheless maintained links across the globe, persuading, begging and coercing his correspondents to send specimens, confirm identifications, do experiments, and so on. Later, after the *Origin of Species* (1859), Darwin carried on writing, now working to get his ideas accepted. Like modern scientists, he worked hard to win over the doubters, flatter his supporters, and neutralise his critics. Darwin's strange life, and his diverse skills, are described in many excellent biographies, including Janet Browne's two-volume *Charles Darwin: Voyaging* (Volume 1, London: Jonathan Cape, 1995 & Pimlico, 1996; Volume 2, London: Pimlico, 2002) and Adrian Desmond and James Moore's *Darwin* (Harmondsworth: Penguin, 1992).
[4] *Charles Darwin's Beagle Diary*, edited by R.L. Keynes (Cambridge: Cambridge University Press, 1988), p. xxiiii.

his experiments and his reading to develop and test his theory. Some
books proved very important. For example on the *Beagle* voyage Darwin
had read Charles Lyell's *Principles of Geology*,[5] which argued that the world
has been formed by the same slowly acting forces that we see today. Weath-
ering and erosion, earthquakes and land movements, can over time produce all the features seen now, including the most gigantic, for example
the Himalayas. The Ganges formed just one of Lyell's examples. He noted
that the river Ganges, when in spate, contains perhaps one part in four
of mud, and from this he calculated that a mass about equal to the Great
Pyramid passes into the Bay of Bengal every 2 days – probably an over-
estimate, but the point is made. As a geologist, Darwin was in a position
to appreciate how Lyell's small but constant changes could add up over
time to phenomena as vast as the Himalayas, or the Ganges delta. Indeed,
when in Valdivia, Chile, Darwin had a remarkable, confirming experience.
A small earthquake hit, and Darwin was impressed. He wrote in his journal: 'A small earthquake like this at once destroys the oldest associations;
the world, the very emblem of all that is solid, moves beneath our feet like
a crust over fluid; one second of time conveys to the mind a strange idea
of insecurity, which hours of reflection would never create.' In the nearby
island town of Concepción, Darwin saw too that the physical equivalent
of his psychological realisation of a changing world was all too evident: 'I
believe this earthquake has done more in degrading or lessening the size
of the island than 100 years of ordinary wear and tear.' It was clear that the
level of land and water had changed. When Captain Fitzroy made the necessary measurements, by looking, for example, at the intertidal molluscs
left high and dry and rotting, and by interviewing local people, he found
that the land had certainly risen by 6 feet.

One vision of nature is that it is beautiful and fine. Darwin, brought
close to nature by the rigours of his youthful voyage, saw in some of his
later reading a different version of nature that he recognised as true. A key
text was a book by the economist Thomas Malthus, *Essay on the Principle
of Population*.[6] Malthus was interested in how human populations were
controlled. With unlimited resources, a population would rise exponen-
tially. Populations rarely behave like that, however. In the case of humans, according to Malthus, the brake on growth is provided by lack of

[5] Three volumes, the first published in 1830, and the others by 1833.
[6] First published as a pamphlet, *Essay on the Population as it Affects the Future Improvement of Society, with remarks on the Speculations of Mr. Godwin, M. Condorcet, and Other Writers*, in 1798, and later, in 1803, expanded into a book, which Malthus continued to develop up to a 6th edition in 1826.

resources (famine), by disease and by war. On reading this, Darwin saw the same process mirrored in nature. Organisms have the potential for prodigious population growth. A pair of birds, for example, might raise six chicks in a year, for several years. A salmon may lay tens of thousands of eggs, of which hundreds may hatch. If just two survive, the population stays level, and as populations normally do stay level, Darwin saw that there must be a tremendous cull in nature. Few survive: those that do are the best able to find food, flee predators, avoid parasitism – in Darwin's terms, they are the best adapted to the current environment. If there is variation amongst a population of one species, and the variation can be inherited, then the survivors will pass on their favourable characteristics. In time, such characteristics will come to dominate the population, but as an environment is constantly changing, so will the population and its inherited characteristics change. Thus both evolution and the origin of adaptedness are explained. Darwin was to spend 20 years fleshing out his mechanism, to be called 'The Theory of Natural Selection'; in 1859, under pressure from colleagues, he finally published it in his book, *The Origin of Species by Means of Natural Selection*.[7] It sold out on the first day.

3.3 The question of time

Compared with the laboratory routines of a research biologist, the major events of evolution occur remarkably slowly. Some aspects of evolution are quick, it is true: a point mutation caused by an incoming gamma ray can take no time at all; the changing frequencies of genes, for example within a population of insects, can sometimes be tracked over a period of months; and microbiologists can observe a population evolve antibiotic resistance in a matter of hours. These processes are evolutionary. Still, the major events of evolution take a long time, and researchers' lifespans are too short to measure them.[8] Consider three examples: the origin of the

[7] *On the Origin of Species by Means of Natural Selection, or, the Preservation of Favoured Races in the Struggle for Life'* (London: Murray).
[8] Palaeontologists have a lot of trouble deciding when something is an 'event'. There are a number of celebrated 'events' in evolution, for example the movement of fish onto dry land, and the extinction of the dinosaurs – but is the highlighting of such 'key events' simply a sign of our parochial human perspective? We have to admit that those first flapping movements by a fish on a muddy shore, and the lingering demise of the dinosaurs, are both part of our own extended history, and so of popular consciousness. Not surprisingly, therefore, palaeontologists are very interested in guaranteeing their objectivity – the separation of their theories from human vanities – and so they search passionately for accurate ways of measuring and of classifying that will remove the element of 'self' from their findings.

dinosaurs, their spell on earth, and their final extinction; or the separation of one population from another by the uprising of a mountain ridge (such as the Rocky Mountains in the USA); or the adaptive radiation of immigrant organisms on recently formed islands (such as Darwin's famous finches of the Galapagos). Here, things happened over tens of thousands, hundreds of thousands, or even hundreds of millions of years. Run a few of these events together, and the time demanded begins to mount up. Indeed, the most basic aspect of evolution is how much time it has taken.

Imagining these lengths of time is not easy. Our basic units are rooted in lengths and times that relate closely to the human dimension: a metre, a second, a month, a year. How do you appreciate time when it stretches so far back there is no measuring instrument, no ancient institution, not even any mountain or ocean that occupies more than a fraction of the prolonged history of the earth? Now, it would not be true to say that of all the people who have ever lived, only modern geologists have been able to conceive of long lengths of time ('Deep Time', as it is sometimes called). There are plenty of examples of people referring to a time deep in the past, and to the events of that distant era. The aboriginal Australians, for example, have as their creation myth a time in the remotest past, the 'Dreamtime' or *Alcheringa*, when the land was transformed from a desolate featureless landscape to a rich terrain with mountains and valleys, rivers and hills, waterholes, animals and plants. The change was produced by 'Creation Ancestors' who emerged from the depths of the earth or from the sky; they made epic journeys overland, creating land formations and living things as they went. This creation took place a long time ago; modern aboriginal Australians do not connect to that time by drawing a time line or by estimating in millions of years when all this happened, but by ritual. Chants, dances and poetry evoke the Dreamtime, commemorate the Ancestors, and retell the story of the creation of the world. Ritual keeps the Dreamtime significant even though it is immeasurably remote.

The geologist, however, gives reality to the concept of Deep Time by measurement. Though the oldest rock so far measured is 3.5 billion years old, and though that time is beyond our imagination, we use our own familiar units – years – to get some intellectual traction on the distant past. We say 'The Cambrian explosion took place 570 million years ago.' We attempt a connection to that time by using the same simple unit that measures our own lifespans (the year). The ability to measure the very ancient lies at the root of our thinking about evolution. For now, with this concept of Deep Time, we can debate the origin of life a billion years ago, and attempt to understand the fossils that come our way. In evolutionary theory, chronology and chronometry are important.

Yet diagrams and visual aids that depict the chronology struggle to avoid distortion. We can see this by strolling through one of those museum galleries that present as a panorama along a wall the march of earth's species towards modernity. The boiling, steamy surface of the earth gives rise to the first cells, or even proto-cells, perhaps 4 billion years ago. Further along you see the emergence of the first multicellular organisms 700 million years ago, and then the arrival of hard-shelled invertebrates a few feet (or 130 million years) later. Then, in quick succession you pace by the trilobites, the first vertebrates, the first land vertebrates to get out onto dry land, the arrival of the birds, and so on. As you walk the illusion is created that the whole of life is lurching forward: as though all bacteria became eukaryotes, or all fish left the sea and became land animals, or all reptiles took to the air. Of course only a few populations made these 'pioneering' changes, the rest remained, so that today we still have bacteria in ponds, fish in the sea, and chimps in the forest. Though an apparently obvious point, we shall refer to this illusion of an inexorable march of evolution when we look at the powerful grip exerted on evolutionary thinking by the concept of progress.

Our own involvement in the time line is ambiguous. We can lay out with fair accuracy the events of life. Yet the museum gallery can hardly be to scale. If it were to scale, then fully half the wall would have to be taken up by empty space, representing a mysterious time span of at least 2 billion years when life was only bacterial. The problem however would be worse at the other end, when the organisms so grandly able to build and enjoy museums – we humans – would be compressed into the tiniest of fonts and almost out of the exhibit, for humans have only been here for a few tens of thousands of years. So for a more realistic metaphor we have to think of 24-hour clocks, with humans occupying the second before midnight. Some popularisers of science use the metaphor of a human arm: if the length of the human arm is taken as the time span of life on earth, with the most modern portions represented by the wrist, the fingers and the finger tips, then *Homo sapiens* would be removed without trace simply by lightly passing a file over the fingernails.

An evolutionary time line drawn on a page, or installed in a museum, is likely to be distorted in order to give enough space to mammals, and to reduce the boredom of visitors pacing past metres of space devoted only to microbes. These time sequences may also be misleading in another way, for they suggest that the organisms drawn and modelled are related to each other, that one form led to another. Yet can one know that one fossil is the direct descendent of another organism living 3, 10 or 15 million years earlier? The biologist and journalist Henry Gee, in his book *Deep Time*,[9] uses this doubt to undermine the assertion that evolutionists can ever know a line of descent. He puts it this way:

> Deep Time is like an endless dark corridor, with no landmarks to give it scale. This darkness is occasionally pierced by a shaft of light from an open door. Peering into the lighted room, we see a tableau of unfamiliar characteristics from the lost past, but we are unable to connect the scene before us with that encountered in any other room in the corridor of time – or with our own time. Deep time is fragmented, something qualitatively different from the richly interwoven tapestry of time afforded by our everyday experience, what I call 'everyday time' or 'ordinary time'.

Gee's point is not only that Deep Time is beyond our imagination (the same could be said of the vast distances of space); it has no inherent structure either. We try to give it structure by anchoring fossils to certain dates, and by suggesting that one fossil is an ancestor of another. Those time lines in museums imply not only that one kind of organism followed another, but that they were related by descent. According to Gee, this is a fallacy. Fossils are rare, and Deep Time is immense. Each fossil is just a pinprick of light in a vast darkness. You can identify and date a fossil, but you cannot know its ancestor, nor its descendents. Gee puts it this way: 'You cannot connect one fossil with another to form a narrative.'

Richard Fortey, in his book *Life: an Unauthorised Biography*,[10] gives another perspective on Deep Time. This time the point of interest is not so much the debatable scientific value of accounts of descent, but the difficulty of measuring the age of the fossils themselves. For as you go back in time, the accuracy diminishes. An event happening around a million years ago can be dated to within a few thousand years. The margin of error in dating something 100 million years ago will have grown to as much as 100 000 years. For an event 400 million years ago, dating it to within half a million years might be manageable. So as you look back in time, the accuracy of pinpointing evolutionary events diminishes.

---

[9] *Deep Time: Cladistics, the Revolution in Evolution* (London: Fourth Estate, 2001).
[10] London: HarperCollins, 1997

The judicious student will always view with caution the more ambitious statements of any evolutionist. Of course, as we saw in Chapter 1, the idea that science of any sort deals in timeless truths is worth challenging. With evolutionary theory, the great lengths of time involved, and the defects and gaps in the fossil record, make for a science where evidence is constantly being propped up by the demands, desires and prior knowledge of the scientists involved. It is an area where disputes over the meaning of data are the rule rather than the exception. Often, the zealous interpretation of scanty material is legitimate and knowledgeable. The eighteenth century French evolutionist Cuvier is still renowned for his boast that from a single molar tooth he could dream up an entire skeleton. Clearly scientific data only becomes valuable in the hands of a skilled, interpretative scientific mind that can fit the data into some greater scheme, or use it to challenge an orthodoxy. Also quite clearly, the evidence can be pushed and pulled too far, being used in support of beliefs that, perhaps much later, come to be seen as quite insupportable. In short, with so problematic a source of information as the fossil record, it is only too easy for the worker to invest too great a meaning in the data. One result of this is that evidence of evolutionary theory is always argued over. Another result is the constant search for new methods that will remove sources of ambiguity, such as the variety of human interpretation. This is the justification for the field of cladistics, which simply measures similarities between organisms, and thus their relatedness. Cladistics avoids discussing what it sees as dubious half-truths about evolutionary descent, and simply concentrates on comparing the relative closeness of different groups.

Throughout this chapter we will see that the arguments of evolution are arguments not only about theory, but also about the meaning of evidence. In our first case study, on Jacques Déprat, we have a situation that, according to your viewpoint, amounts either to fraud, or to visionary work years ahead of its time.

3.4 The case of Jaques Déprat

Jaques Déprat, a French geologist, died in 1935. At the time of his death, he was known in scientific circles as a fraud: his crime, so it was alleged, was that he had made up his results. Specifically, unusual fossils that he had 'found' when prospecting in Vietnam were actually European in origin, and had been fraudulently included in his Indo-Chinese collection to make a stir. This at least was what an eminent Paris science committee announced on 4 June 1919. Déprat counter-attacked: he really had discovered these fossils in Vietnam. Yet in a contest between the forces of a grand, conservative committee, and a single scientist who had found something remarkable, the committee was always going to win. By the end of the year, Déprat was ousted. Expelled from the French Geological Society and sacked from his university post, his career was over. As it happened, he fashioned himself another job as a novelist, and quite a successful one. He wrote a number of political and autobiographical works under the pseudonym Herbert Wild, one of which, *Les Chiens Aboient* (*The Baying Hounds*, published in 1926), caused a small scandal by containing, in disguise, an attack on the men who had brought him down. This career was going well when a second and final tragedy struck: Déprat's geology had continued in the form of high ascents in the Pyrenees, and it was on one such expedition, near Aragon, that he fell and died.

All this would have been a near-forgotten detail of France's glorious scientific heritage, were it not for some diligent reassessment of Déprat in the late 1980s. The research brought to light how brilliantly successful Déprat had been as a geologist, and suggested that he had stumbled across an idea 50 years too soon: The Theory of Tectonic Plates. This particular conclusion has been contested by other contemporary palaeontologists, for example Richard Fortey in his book *Trilobite!*;[11] but Déprat has at the very least emerged as an esteemed member of the French geological community at the start of the twentieth century, rather than as a committed charlatan. He produced his first geological paper at the age of 19, before gaining his doctorate. By the age of 28, Déprat had led a whole series of successful – and dangerous – missions surveying the volcanoes of Sardinia. His skills became known to the professors of French science, and he took up a posting in French Indo-China. Déprat was, it seems, a brave explorer, and between 1909 and 1919 he researched an area in the north of Vietnam, where rough travelling conditions were made consistently more troublesome by bands of robbers. Even with these problems, he managed to produce a stream of papers judged to be of exceptional quality. The result was that Déprat was declared to be amongst the very finest geologists of his generation.

Among the fossils Déprat dug up were some trilobites. They were more than interesting, for they had very strong affinities with European

---

[11] *Trilobite! Eyewitness to Evolution* (London: HarperCollins, 2000).

trilobites, and as a result Déprat was able to calibrate the ages of the Vietnamese formations he was working with – an important achievement. However, as described by Michel Durand-Delga, the geologist who re-visited and reconstructed this intriguing episode, the drama really be-gan in February 1917. According to Durand-Delga, colleagues with a grudge against Déprat started a whispering campaign against him and had his fossils re-examined. They announced that the fossils were entirely European and alleged that Déprat had brought them out from Europe and inserted them into his Indo-Chinese collection. Déprat's problem was that he had no way of explaining why so European-looking a set of fossils could be in Indo-China. His finds were unique and, given the conditions, more than usually hard to replicate; they were also plainly European fossil types; and perhaps more seriously, there was no theory, other than fraud, that could explain the presence of European fossils in Indo-China. Hence the expulsion of Jaques Déprat from French academic geology in 1919.

Doubts about the justice of this dismissal emerged in the early 1990s, when Durand-Delga presented his research.[12] The defence of Déprat partly consisted of an analysis of the animosity of his colleagues, with the implication of doubts about their even-handedness, but there was also a scientific defence. For with the development of plate tectonic theory in the 1960s, it had become possible to understand the reasons for similar-ities between fossils in Vietnam and Europe. Plate tectonics allows for movement of giant surface plates across the surface of the globe, and, 400–500 million years ago, the land masses of Southern Europe and South-East Asia may have shared more similarities of latitude than they do now. Trilobite fossils extended eastwards from Europe around a Gondwana palaeocontinent, following the comparatively cool, circum-polar shelf seas. Déprat's modern-day defence goes like this: no wonder Déprat's fossils looked European, in a sense they were, but they were carried to Asia as part of Gondwana, not by skulduggery. According to Durand-Delga, Déprat was groping towards such a solution in his very last paper, but the scientific world was not yet able to see continents as mobile, especially if the proposal came from someone suspected of fraud.

Does the posthumous reappraisal of Déprat close the episode? The ar-rival of plate tectonic theory makes Déprat's claims possibly true, rather than certainly false, and as this is a pretty story, with a downtrodden

---

[12] 'L'affaire Déprat', *Travaux du Comité de l'Histoire de Géologie (Cofrhigeo)*, **4** (1990), 117–212; 'L'affaire Déprat', l'honneur retrouvé d'un géologue', *La Recherche*, No. 237 (1991), 1342–6.

man victimised by an overbearing and envious committee, we certainly feel tempted to see Déprat not as a cheat, but as a hapless victim of his own vision, a man-before-his-time genius. However, this, again, might be reading too much into the evidence. For, in a paper in 1994, another dissenting voice was heard, from another French palaeontologist, J.-L. Henry, publishing far away from the controversy in the Australian journal *Alcheringa*.[13] Henry states flatly towards the beginning of his paper 'The trilobites of the "affaire Deprat"': 'In my opinion, all these trilobites really come from Bohemia, where such forms are well known.' Henry directs his own knowledge of trilobites at the case, ignoring the human aspect of the story. His is a sober analysis of what we know of the actual trilobites found by Déprat. The fossil identification is not under dispute; they are known today as *Deanaspis goldfussi*, *Dionide formosa* and *Dalmanitina socialis* – but could such fossils plausibly be found in the kind of strata claimed by Déprat, and does our current knowledge of the biology of trilobites support Déprat, or not? On both counts, Henry inclines to the negative. For example *Dionide formosa* is a deep-water trilobite, and restricted to the upper Ordovician, a rock age that Henry claims as absent from the area Déprat worked in. The earlier, lower Ordovician, does exist in the area, but in intertidal, shallow water conditions. Henry notes 'The deception seems evident here; it would be judicious to establish whether the 3-m-thick calchist bed described and illustrated by Déprat is really present at the locality indicated.' He claims that while there are similarities between Gondwana Ordovician Asian trilobites and their European cousins, the similarity is only at the generic level. It remains the case to this day that the species described by Déprat, commonly found in geological collections, have always been found in Europe, never in Indo-China.

The argument is likely to continue. On one hand we have Déprat's serious scholarly career and the fact that plate tectonics could in principle account for those species turning up in Vietnam. On the other hand, if we look in greater detail at the actual fact of those species, we not only find that there have been no subsequent Indo-Chinese 'sightings', but also learn that their particular habitats are not represented in the areas that Déprat surveyed. As with a scientific hypothesis, there is refutation and there is corroboration. Some of the evidence is rather general, and we may have doubts about a motive for fraud. Some is more precise, as examplified by our knowledge of trilobite habitat, and in the background is the theory

[13] *Alcheringa*, **18**, 359–62.

of plate tectonics, which rules out the general impossibility of European forms also occurring in Vietnam. Further evidence on "L'affaire Déprat" will emerge in time, and only properly appraised fieldwork can ever settle the matter.[14]

3.5 The myth of the coelacanth

Although a fossil can carry a remarkable amount of information, it is still a piece of rock rather than a living thing. Hard parts tend to be preserved, not soft parts. In the case of vertebrates, for example, the animal's physiology and behaviour are much more difficult to interpret than the relation of one bone to another. It is the skill of the palaeontologist to fill in these gaps in the evidence; but when does a fail-safe deduction become an educated guess, and when does an educated guess become a fantasy?

The easy way in which one can slide from one to the other is made plain by the story of the coelacanth *Latimeria chalumnae*, a rare deep-sea fish found around the Comoro Islands off East Africa. The significance of *Latimeria* has rested on the belief that it is uniquely able to tell us about the past, and in particular its important ancestors. Those ancestral coelacanths were extinct 80 million years ago, and were, up until recently, considered to be the direct descendants of the fish that crawled onto land and so pioneered the so-called 'invasion of the land'. The living coelacanth combines a simple, important story (its possible relationship with the first amphibian) with elusiveness (only a few hundred have been fished up in 60 years), and so has provoked a great deal of speculation over the years. The root of the story lies in that moment when a fish hauled itself out of the water and found itself able to function on land. That fish, and others like it, evolved into the commonly accepted first tetrapod, *Icthyostega*, the ancestor of all terrestrial vertebrates, including, of course, *Homo sapiens*. The question asked by palaeontologists has been: what was that fish that came onto the land, during the Devonian period, around 380 million years ago, and how did it cope? This fish must have been a lobefin, that is, a kind of fish where the fins project from stumpy but moveable lobes. To a degree, these lobes resemble primitive limbs with an internal skeleton, and this is why the lobefins are the obvious ancestor of

---

[14] A book has already been published on the controversy. Roger Osborne's *The Déprat Affair: Ambition, Revenge and Deceit in French Indo-China* (London: Jonathan Cape, 1999) is both entertaining and scholarly and reveals the truth of the editors' maxim for would-be science writers: keep it personal. Evolution is a branch of biology where scientists have been particularly fond of writing books, both professional and popular.

the terrestrial pioneer. Yet there are three kinds of lobefins: the lungfishes, the rhipidistians, and the coelacanths, and which of these might be the adventurous group duly provoked much debate. The lungfishes, still around today, were rejected: their skull structure appeared to be too specialised. The rhipidistian group were considered more likely candidates because they have a similar head structure to that of the early tetrapods, and because coelacanths were thought to be direct descendants of rhipidistians, the coelacanth became the closest living relative of tetrapods. The features of the coelacanth fossils, including pronounced limb-like fleshy fins seemed to be just a short step from the tetrapod pattern, and so for many years the coelacanth became established as the real ancestor.

Not surprisingly, intense interest is going to be focused on a fossil considered to reveal truths about the invasion of land. Yet information a palaeontologist would dearly love to have, such as how the fish paddled its fins, whether those lobes could lift the fish bodily from the ground unaided by buoyancy, and general facts about its physiology, are hard to glean from a fossil. The temptation was always there to 'read into' the fossils, information that springs simply from the imagination of the scientist. Subsequent events in the story of the coelacanth show how this happened.

Astonishingly, in 1938, a living coelacanth, weighing 127 lbs, was caught off the east coast of South Africa. A local museum curator, Miss M. Courtenay-Latimer had let it be known that she would be interested in seeing any unusual fish brought out of the sea, and she was duly invited onto the *Nerine*, a trawler just into harbour, to examine a catch. The large blue fish that caught the attention of Courtenay-Latimer was loaded into a taxi and transported back to the museum. She realised this was an important and unusual fish, and wrapped it in a formalin-soaked cloth to slow the rot. Nonetheless, by the time Professor James Smith (a South African chemist and icthyologist) had arrived, the insides of the fish had passed tolerable limits, leaving only the outer skin and the head. They were enough to allow the announcement that a living coelacanth had been found, a representative of a group thought to be extinct for more than 80 million years. More dramatically, and more newsworthy, was the fact that this fish was declared a living descendant of the group of fish that had left the water, and in time, had evolved into the first amphibian. For science, here was an animal that could vastly supplement the information from the fossil record. The respiring, moving and excreting coelacanth might contain in its behaviour and physiology memories of its pioneering past, clues that would give palaeontologists a better under-standing of the fish that first crawled onto land. Thus, after the first coela-canth was found, a £100 reward was offered for the next. In those days this was a lot of money, especially as it was directed at East African fish-ermen, but it took 12 years for another coelacanth to turn up. This time it was found in the French Comoro Islands, a small volcanic group between Africa and Madagascar. Actually, the fish was reasonably well known by local fishermen, but only when the huge reward became known to them, was one kept and sent to the scientists.[15]

Though many specimens have now been studied, the coelacanth has not answered every question about our early amphibious ancestors. And in that failure lie some interesting philosophical points, to which we can now turn. In the first forty years after the discovery of the living coelacanth, it was uncontroversial to see the fish as the direct descendant of the ancestor that first climbed onto dry land. It was only too easy to project the biology of the living coelacanth back onto the actual ancestor, seeing every aspect of coelacanth biology as a 'relic' from the past. Practi-cal issues produced distortions too. Coelacanths live about 200 m down, and do not survive being brought up to the surface. This makes it difficult to examine, for example, how a coelacanth swims, a behaviour that might give information about how an ancestor moved around the shoreline. In general, the research done was influenced by the basic assumption that the biology of the coelacanth would reveal to us the biology of its pioneering ancestor. Rather than being seen as a modern organism, with its own set of adaptations, the coelacanth was seen as a view back to the past.

One of the puzzles surrounding the coelacanth was the question of how it had survived at all. It is a commonplace amongst biologists to believe that when an organism has apparently not evolved much over a great length of time, then that organism has been in a secure, unchang-ing environment, under no pressure to evolve. Difficult environments like the deep ocean or hot springs feature hardly evolving organisms; they are occupying a specialist niche, the niche is stable, and there are few

---

[15] Professor James Smith was lent a Dakota aeroplane to make the journey to the Comoros, because the South African prime minister had been told that an important scientific discovery was in the offing, and could bring the country prestige. As is often the case, the pious idea that science knows no national boundaries came under strain. After the Professor had flown out the *Latimeria*, without either landing papers or permission from the French, the Paris government made very sure that they controlled all subsequent coelacanth research. The story is told in *A Fish Caught in Time* (London: Fourth Estate, 1999) by Samantha Weinberg. For exhaustive detail on the fish, rather than the personalities, consult Peter Forey's *History of the Coelacanth Fishes* (P.L. Forey, London: Chapman & Hall, 1998).

competitors. If the coelacanth is described as 'a living fossil', then it is a short step to imagining that it has not evolved, and must, therefore, have been occupying a specialist niche. This is how the story of the caves arose, back in the 1970s. The coelacanths of the Comoros occupy caves, deep down at the base of the volcanic islands. The idea sprung up that these caves are brackish, filling with rainwater running off the sides of the volcanoes. Brackish water, as is well known, is a 'difficult' environment; if coelacanths had mastered it, then perhaps they would suffer little or no competition – which could explain their sluggish evolution. Even better, the idea fits in with something that is well known about the Devonian period. The animal that came out onto dry land came out from a lake or river, not the ocean. A neat connection is made between the brackish coelacanth, and its freshwater ancestor.

Yet there was never any evidence that these caves were brackish, nor that coelacanths are adapted to freshwater. Moreover, the islands of Comoro are only a few million years old, raising the question of what the coelacanths were doing before then. In fact, the story of the brackish coelacanth, so convenient to its status as a living fossil, died a death when new ideas emerged over the ancestry of the first amphibian. The coelacanths had been assumed to be direct descendants of the ancestor, but in the 1970s the lungfishes, previously dismissed as the main candidate, were re-analysed, and declared to be more closely related to the first amphibian. The effect of this on coelacanth research has been quite striking, for suddenly it could no longer be taken for granted that the coelacanth is the closest living analogue to the fish that flopped onto dry land. This meant that the biological straightjacket that had been imposed on *Latimeria*, the belief system that saw the fish as a living fossil rather than an animal with its own suite of adaptations, was removed. The myth of the coelacanth weakened, and certain theories faded. The neat idea that the coelacanth lived in brackish conditions, as befitting an animal with fresh water ancestors, was forgotten.

As the myth of the coelacanth lost its grip, so research into the modernity of the fish began to assert itself. How can you distinguish an ancient adaptation from a modern one? The former might tell us something about the first land vertebrate, the latter tells us about the living coelacanth. One mystery resolved was the acute problem that coelacanths face when brought into shallow water: a problem that means that they cannot be observed for more than an hour or so in an aquarium or surface net. Compared with other fishes, a coelacanth has an unusually small gill surface area to body mass ratio. When it comes up to the relatively warm waters of the surface, and thus experiences water with a lower partial pressure of oxygen, its gills cannot absorb enough oxygen for its needs, and so it suffocates. As research continues into this and other characteristics of the coelacanth we will come to see the animal as not simply a repository of ancient-world features, but as an animal perfectly adapted to its modern world, with its own unique set of design solutions. It will not be easy to distinguish the two types of feature. For example it has been found that the coelacanth has a tapetum, a reflective zone behind the eye that aids vision in low light conditions: is this a modernism? Submersible cameras reveal that *Latimeria* swims with a doggy-paddle that is reminiscent of the limb co-ordination of land vertebrates: is this an authentic clue to the past?

The coelacanth story has a sequel. In 1998, a coelacanth was fished up, not in the Comoro Islands, but 10000 miles away, off Sulawesi, in Indonesia. This discovery is likely to provide one more blow against the coelacanth myth. For even if we accept for a moment the idea that the Comoran population is a 'lost world' population, stranded outside its time, but hanging on in caves at the base of volcanic islands, the idea that there are two such populations, separated by an ocean, defies belief. An alternative is that *Latimeria* is native throughout the Indian Ocean, and simply has not been noticed outside the Comoros. The Indian Ocean is a remote area, and it turns out that the fish has always been reasonably familiar to Indonesian fishermen, who call it *Raja Laut*, but the fish was not considered good food, and no one had heard of the 50-year-old coelacanth industry operating out of the Comoros. No doubt that will now change, and it is hoped that the Indonesian discovery will show the coelacanth to be commoner than was thought. Its conservation would be easier if it was a less-celebrated icon of science.

3.6 Life and the illusion of progress

Is evolution, by definition, also progress? The background to the issue is simple. Evolutionary science has shown not only that life has changed, but also that the change has a direction. The planet has seen the successive emergence of eukaryotes, multicellular organisms, nervous systems of greater and greater complexity, and the invasion of land and air by terrestrial vertebrates. Our modern horses, which run so speedily around a racetrack, are traditionally described as having evolved from very much smaller creatures, perhaps the size of fox terriers, which lived in the woods, and ate berries. Comparing the modern or latest form with the earlier, surely we see progress?

One reason for the direction of evolution comes simply from the fact that mutation is random, and natural selection consists of an environment operating on the variation (and differential fitness) presented by a population. For a lineage to evolve backwards, a genome would have to mutate back to a past state, with the environment itself following suit. Bearing in mind the overwhelmingly complex relations between members of a population and the environment, the precise reversal of evolution can be ruled out. Even the idea that evolution shows trends needs cautious treatment. Such trends (for example 'getting bigger over time') can sometimes crop us as a result of sampling error. Today's only surviving horse, the genus *Equus*, is much bigger than the early representative of the horse family *Hyracotherium*. The small has become big, the slow fast. As we view the horse as in essence a fleet-footed animal, and risk fortunes on that quality, we might well see the road from *Hyracotherium* to *Equus* as sure progress. Yet if you take into account all horses in the fossil record, no such trend emerges. It is just as true to say that some lineages of horses, for example, got smaller over time. It all depends on what fossils you have found, and how you do the statistics. Progress is, therefore, hard to discern in practice and difficult to justify in theory. If we say that evolution shows progress, are we saying something scientific? For most scientists the concept of evolutionary progress is empty if it relies on the incorporation of the supernatural, such as an external God, or inner 'life forces'. Equally, if the concept is simply a human value applied to nature, then again it is suspect.

For a system to be showing progress, it must be getting better. This begs the question: what is better? Sometimes there is little argument. We might see progress in the punctuality of a railway network, in the academic record of a student, and in the human rights record of a particular country. In these cases we go beyond merely acknowledging a change: we declare that things have changed for the better. We can identify the progress because there are agreed criteria. In the case of a railway network, progress might well be judged on punctuality. The manner in which a country treats its prisoners might change, and could be considered a valid measure of its progress towards becoming a civilised nation. Still, even in these relatively straightforward cases, we can admit that some will dissent. Passengers of a railway company may hold that punctuality, not profitability, is the most important criterion of progress; the shareholders could have the opposite view. Finding a criterion to judge progress in evolution is even more fraught.

For example we might try to argue that evolution has been progressive because a rather recent manifestation of its processes, the arrival of *Homo sapiens*, has produced the most intelligent, the most communicative and the most creative of all organisms, ever. We have the most advanced brain; we have language; we have consciousness. Thus, though we are a new arrival on planet earth, we at least are the most impressive of its inhabitants; what we may lack in seniority, we make up for in talent. Yet it is not hard to see that this argument suffers from an outrageous anthropomorphism. It is we people looking at the whole of life as a forerunner and preparation for our arrival. We look at the attributes that we see as particularly human – intelligence, language, versatility – and rate the accomplishments of other species in terms of how close they have got to the human norm. Some organisms are more intelligent, communicative and versatile than others, and we can rank them accordingly, but always with the pleasing thought that we occupy, by definition, an unassailable position. The argument is not one that can be sustained scientifically, because this way of judging progress is so obviously a value judgement: that consciousness is best.

There are more sophisticated versions of the meaning of progress described above. You can strip away all talk of humans, and simply study one 'objective' s characteristic, which can be shown to have increased over time: complexity for example. We can then begin the task of seeing the mammal as progress over the maggot by simply pointing to its increased complexity. You need in turn a yardstick for measuring complexity; some aspect of the nervous system might prove suitable. In the case of the vertebrates, features of the skeleton are often used for evaluating complexity. These measurements can then depict the story of the vertebrates as also the story of progress. Judged by their complexity, there is progress from fish to amphibians to mammals. Within the mammals too, the greatest complexity is shown by the latest arrivals, but that is not the same as saying that the mammals as a group show progress over time.

Parasites give an interesting sidelight on the utility of complexity as a measure of progress. For in groups where there are both free-living and parasitic members (such as the nematodes), it has often been noted that the parasitic way of life involves a loss of structure. For example locomotory or digestive organs may be atrophied. Yet though these diminutions may be especially obvious, it is not likely that they can be used as a sign of loss of complexity. For a start, there may be compensating developments, such as the hypertrophy of the ovaries, or adaptation of mouthparts into well-developed hooks or suckers. The life cycle too might show an increase in complexity, so that the many larval stages are parasites, perhaps of organisms left untouched by the adult. Finally, in judging the complexity of a parasite, relative to a free-living cousin, it would be hard to know how to balance the changes in anatomy and behaviour against those in genetics and biochemistry.

Organism genomes have often been used in an attempt to find an objective measure of complexity. However we define genes, and however much we might argue over the manner in which a genome makes its contribution to the final form of an individual organism, it cannot be denied that the genome is a fundamental aspect of an organism. With so much work under way on mapping and sequencing the DNA of different organisms, why not simply measure the genome and measure the progress of organisms according to the size of the genome? Here though, there are some teething problems to sort out: a newt has five times more DNA than a human, a distortion that is corrected when the actual number of coding base pairs is calculated. When genome size is worked out using this measure, humans score better than newts, and vertebrates better than invertebrates. However, the number of coding base pairs may give a rather imperfect indication of the complexity of an organism. The work of the Human Genome Project, and allied research, suggests that it is the control and expression of genes, rather than their number, that tells the difference between a yeast, a mouse and a human.

Of course, in one sense, this still resembles the earlier, highly anthropomorphic version already discussed in this section: humans come out best. This time, however, the idea of progress relies on a concept – complexity – that may be measurable. Here we run into a problem only too common in science. If we set out to measure something, and find we can do so quite easily, generating strings of data in the process, we may, nevertheless, be engaged on an utterly trivial task: in science, measuring is not enough.

There are other problems with the attempt to use complexity to define and track progress. The first argument is that the charge of anthropomorphism still lurks: is not complexity a quality that we judge important simply because our own endowment is so impressive? In other words, it is an important aspect of life, but it may not be appropriate for evaluating progress. In terms of evolution, the most fundamentally important aspect is certainly the fit of an organism to its environment. On that account, the limpet is as well adapted as a marmoset or an oak tree or a human. Each organism has its own set of adaptations: limpets have their glue, humans their intelligence. Both organisms are well adapted. You cannot rank organisms according to adaptedness, for all surviving species are by definition well adapted. Within a population, of course, some individuals will be better adapted: this is the root of natural selection, as we shall see shortly, but the concept of progress is used to rank species, not individuals amongst a population. Population geneticists measure fitness differences amongst genes, and describe the 'progress' of genes through a population; but that progress usually depends on the relation between the gene and the environment, not on any inherent quality alone (such as size or complexity). It seems, then, that the use of complexity to measure progress is little more than a back-door attempt to justify a human-centred view of the evolutionary story of life.

The second argument attacking the concept of progress is one that examines neither the complexity criterion, nor the way it is measured. The attack is on the way the measurements are compiled: the statistics. In this argument, progress is an artefact of the statistics, or rather of the way that statistics can be used to tell a variety of stories. That is to say, the impression of increasing complexity is itself simply an impression. Most organisms are bacteria, and while in every age, new and more complex organisms have evolved, the whole of life has not lurched in one direction. The seas are still filled with jellyfish, and the ponds with protists. These modern representatives of ancient phyla have their modern adaptations, but, if one was driven to classify them on a scale of complexity, and decided to judge these organisms on the basis of their nervous systems, they are still what we would call 'simple'. It would be a crude judgement, as any protozoologist will tell you: no protozoologist sees *Paramecium* as a 'simple', or worse, a 'lower' organism. In any event, to see life as showing progress is to be in danger of simply ignoring the life that lies far from us on the evolutionary scale, but is just as well adapted as we are. Progress, in short, is a value judgement.

3.7 Proximate and ultimate causes in biology

It was a problem for Darwin that he had no theory of inheritance. Though he saw that variation could be the generator of evolution so long as surviving variation could be inherited, he had no knowledge of genes. His own attempt to speculate on inheritance is today judged a failure. The transformation of Darwinism by genetics took place in 1930, and led to the focus of discussion of evolution being focused as much on genes as on whole organisms. This modification, called neo-Darwinism, still has the same basic structure:

1 Organisms have the potential to produce more offspring than the environment can support.
2 Organisms of the same population differ one from another – variation – and some of this variation can be inherited.
3 Because of this variation, some members of the population are by chance better adapted to the environment.
4 These organisms have better odds of surviving and reproducing than others in the population, and so pass on the favourable characteristics.
5 The result is evolution.

This scheme forms the centre of Darwin's achievement – and can be described as the core theory of biology. Yet there is nothing wrong with maintaining a critical attitude to Darwinism: in spite of its great authority, it is only a scientific theory, and, in fact, one of the pleasures of evolutionary theory is its ability to generate debate. To take one common dispute, there is an argument over the extent to which natural selection accounts for evolution. Darwin's own views on the power of natural selection varied over time; under the pressure of criticism he came to allow other forces also, so that by the time his 6th edition of *The Origin of Species* was published, his claims for the powers of natural selection were somewhat weakened (today it is the 1st, more radical, edition, that is in print, not the 6th). Later in this chapter I will describe the way in which the dispute has rumbled on in modern biology, with debates over the significance of random drift and neutralism, processes seen by modern evolutionists as complementary to natural selection.

Though there are skirmishes about the details of the role of natural selection in nature, the vast majority of biologists see it as essentially correct. For the present, then, let us pursue its features a little more fully, using a concept put forward by the great evolutionist Ernst Mayr. He argues in his book *Towards a New Philosophy of Biology* that two parallel strands exist within biology.[16] One is what we might call functional biology, the other is evolutionary biology. The first is concerned with the here and now of biology. It asks questions such as: what protein does this gene code for, how do pigeons find their way home? The second area concerns itself with

---

[16] *Towards a New Philosophy of Biology: Observations of an Evolutionist* (Cambridge, Mass.: Harvard University Press, 1988).

the evolution of these phenomena, and asks different questions: why has this feature evolved, what is its history? Mayr's point is that you need both strands of biology, working co-operatively. His example of their complementarity, rather than their opposition, is illustrated by the concepts of proximate and ultimate causation, a distinction championed by Mayr. The idea works as in the example described below.

Consider the migration of the Monarch butterfly, *Danaus plexippus*. The migratory journey runs 3000 km from Mexico to Canada, is achieved in 6 months, and involves several stages, each accomplished by one generation. It is a remarkable feat of endurance. All sorts of biological enquiries could be asked about the physiology of the butterfly, such as how often it feeds and how far it can fly in a single day. The Monarchs overwinter in a few Mexican forests, where the trees are literally covered by millions of barely twitching butterflies. A question can, therefore, be asked: what starts them on their journey? According to Mayr, we can have two types of explanation for this, a proximate explanation, and an ultimate explanation. The rising temperature is the proximate explanation. As the forests warm, the Monarch's metabolism fires up, and a cascade of events follows. Clearly, this is the here and now of biology, well investigated by reductionist approaches. However, something is missing, and here we need the ultimate explanation. This considers the reason why Monarchs migrate. Over tens of thousands, or millions of years, Monarchs have evolved a pattern of behaviour that carries them north to the feeding grounds of the Canadian summer. That evolutionary history is as much an explanation for the Monarchs flying north out of the Mexican forests, as the fact that the warming temperatures bring the Monarchs out of hibernation. Throughout biology, the two forms of explanation, proximate and ultimate, coexist. You are likely to be more aware of the former, because the bio-molecular explanations that so dominate biology are themselves so dominated by the links connecting one reaction to another. Yet even in molecular genetics the question of why a feature has evolved – its purpose – can always be asked. Indeed, for Mayr, the biological significance of a chemical reaction within an organism will always lie partly in its contribution to the evolution of that organism.

3.8 The concept of adaptation – and its teleology

Because adaptation is a central concept of Darwinism, it is important to be clear about its precise meaning. For Darwin, as for naturalists before and since, the 'fit' of organisms to their environment is always remarkable.

As every organism fits its environment, examples always reflect the preferences of whoever is writing. For my purposes, I will mention the zebra, and one of its familiar features: the slow swish-swish of the tail, as it attempts to keep the flies away. We call the tail an adaptation, and see its adaptedness in its efficient, life-saving ability to limit the danger of parasites.

You do not have to be a Darwinian to see that it is no chance that the zebra has a fly whisk of a tail. The tail is designed for a job: that of keeping the flies down. Darwin, in formulating his theory of natural selection, knew that his theory must tackle head-on this fundamental aspect of nature: its adaptedness. He had to show that all those superb examples of adaptedness are explained by his theory. In the case of the zebra, it seems likely that as the horse family evolved on the plains, any variation that enhanced the ability of a tail to stop flies settling would be selected.

Such arguments are the staple of Darwinism. In them, two ideas have been run together: natural selection and adaptation. Some 200 years ago, before Darwin, adaptation – the fit of organisms to nature – was not synonymous with natural selection, but now it is. Philosophers and biologists, therefore, declare that if you claim something is an adaptation, you claim at the same time that it is there because of natural selection. To discuss this further, and especially the interesting consequences that it has for what we understand by biological explanation, we must turn to an obscure, but important branch of the philosophy of science, teleology.

When we say something is an adaptation, we give an explanation of how that characteristic came about: the characteristic is there because of what it does. In the case of the zebra's tail, it is there (or at least it has that particular muscularity, mobility and bushiness) because it helps to get rid of flies. With an explanation put as baldly as that, it is not clear what makes it any different from one put forward by someone who believes that a benevolent God gave the zebra its tail. For both the Darwinian and the Creationist, the existence of the tail is explained by its purpose.

The Darwinian, however, has found a way of linking design to existence that does not involve the supernatural. That way, of course, is natural selection. Over hundreds of generations, slight inherited variations that give some members of a population an advantage over others will, in time, come to dominate. The advantage lies in superior adaptedness; the final effect is to achieve this remarkable appearance of design that so marks nature. Yet, in a real way, if something is an adaptation, then its existence is explained by its contribution. For, according to Darwinism, if it did not make a contribution, it would not be there at all.

For a scientist, there is something remarkable about this style of explanation. It does not seem to fall into the normal pattern at all. We can see this by considering a few more examples, such as the planets of our solar system. The moon lies in an orbit on average 384 000 km from the earth. It completes one orbit every 27.32 days and it spins on its own axis in such a way that one side is permanently faced away from us. As a result, each side has a 2-week day, and a 2-week night. During the lunar night temperatures fall to −170 °C; during the lunar day temperatures reach 110 °C. All this can be explained by the laws of physics: the orbit is a matter of inertia, the temperature is a matter of illumination. We would consider strange someone who explained the orbit of the moon in terms of its usefulness in providing tides, and therefore, a diverse littoral ecology on a rocky coast. We do not see any purpose at all in the moon, its orbit or its day length, and nor do we need to when looking for explanations for its behaviour.

Explanations that invoke a future event as a cause are called teleological (from the Greek *telos*, or tail). The Greek philosopher Aristotle is normally considered to be the originator of the idea that nature is teleological. He considered 'causes' (the 'because' of an explanation) to be of various types. For example the material cause of a house is the bricks and mortar. Another type of cause that interested Aristotle was the 'final' cause. In the case of the house, the final cause is its function: it is a place for living in, and someone intends to live in it. In a very clear sense, the house exists not simply because of bricks and mortar, but because of its function, or end. The teleological cause of the house is the purpose that it will fulfil.

In science, teleological arguments are unpopular for two reasons. Firstly, they have long been associated with the supernatural: that such arguments appeal to obscure 'inner tendencies' of things, or to a God that wishes the world to be well designed. Secondly, and following on from this, the favoured and successful paradigm for the physical sciences has been the causal explanation, which invokes how one physical cause impacts on another, and then another: in biology a good example would be the metabolic diagrams you see on posters up on the walls of laboratories.

Yet in biology we talk in a teleological way all the time. Consider the fox that is chasing a rabbit. How can this be explained? Of course there is much to be said of neuronal pathways, muscular contractions and the production of ATP. Yet the explanation surely must include the prospect of the meal ahead. The fox behaves as if it has a goal, and swerves and accelerates accordingly. This is not to say that the fox is thinking along the same lines as we do when dropping in at a sandwich bar; but undeniably the final end – the *telos* – must be part of our explanation for its behaviour.

I said earlier that an adaptation is by definition the result of natural selection. If we explore this further, we see that an adaptation is something that has a function. Thus, the valves of the heart have the function of causing the blood to flow in one direction only, and we explain the existence of the valves by referring to their function, which is the unidirectional flow of blood. This is a teleological explanation because it takes the form of explaining the existence and operation of something by referring to is primary biological purpose; in short, a way of explaining the existence of a heart valve is to refer to its purpose. However, a non-teleological explanation is possible too. Valves can also be explained by the chemical pathways that led to their development, and the rhythmic behaviour of the valves can also be explained by referring to hydrodynamics. For a physical explanation, that might suffice, but to neglect the purpose of the valves in explaining their existence would be to neglect their Darwinian explanation, and thus to miss out a vast part of biology. Darwinism pushes teleological explanation into biology by way of natural selection. According to that theory, every organismic characteristic that we see today is adapted to the environment. The valves are an adaptation of the heart to make it more efficient: they have been selected for because they cause one-way flow, not because they make noises.[17] So a teleological explanation is appropriate: indeed, an adaptive explanation is a teleological explanation. There is, however, no modern equivalent in physics or chemistry, a distinction that supports the claim that biology is distinct from the physical sciences.

No wonder teleology exists in biology. Look at this explanation of the mosquito biting a human hand, and reflect on what is missing:

- Molecules from a human trigger events in the mosquito brain.
- Flying behaviour is controlled by incoming sensory information (amongst other things).
- The heat and the proximity of the skin causes the mosquito to sink its mouthparts deep into a capillary.
- Blood flows into the mosquito.

---

[17] As it happens, heart noises are useful. Babies at the breast like them, and they are important for doctors with stethoscopes, and for their patients, but this kind of usefulness does not explain the existence of heart valves.

There is nothing teleological about this description, nor is there any-thing incorrect. Yet something important, something biological, has been stripped out. There is no concession made to the fact that the mosquito has a goal – feeding – and governs its behaviour accordingly (this does not mean that the mosquito is thinking 'blood …'). When discussing adaptation in organisms, such as the mosquito's mouthparts, we compre-hend very clearly that the existence and form of these exquisitely crafted components really can be explained by the fact they contribute to the mosquito's survival, and have done so throughout the evolution of the mosquito. Compare again this state of affairs with the moon: the moon has no function, and nor does its orbit; both are results of the laws of physics, operating in a normal causal fashion, and no other explanation would be appropriate in science.

3.9 The scope of natural selection[18]

At the end of the introduction of *The Origin of Species*, referring to the evolution of life, Darwin wrote: 'I am convinced that natural selection has been the main but not exclusive means of modification.' It is a thought-provoking statement, and a prescient one too. For a great deal of evolutionary debate, ever since Darwin, has concerned the level of influ-ence of natural selection on life, including humans. The question goes like this: in principle is it true that every feature of life, including the particu-larities of human consciousness, is the result of a history of genetic dif-ference and a resultant differential survival? Those biologists who want to explore the history of evolution as a process where natural selection is a contributory or major factor, but not the sole factor, look for other (non-supernatural) causes. Darwin's comment suggests he was minded to consider these other causes too. He certainly entertained a theory well sup-ported in his time: the modification of organs by use or lack of use, and the subsequent inheritance of those changes. We ourselves see each day the

---

[18] This long section summarises a range of arguments surrounding the pervasiveness of natural selection. The diverse ideas I deal with all have some bearing on the debate that Darwin himself started, namely, the degree to which natural selection may be considered responsible for the organic features that we see around us. This is a central issue in biology, with important philosophical implications, for the more successfully biologists reveal the scope of natural selection in the natural world, the more likely it is that biologists and philosophers will consider that natural selection too is important in understanding the human mind. The range of possible topics is extremely wide; inevitably I have been selective. Some key terms used here are epigenetics, random drift, neutral theory, nearly neutral theory, molecular phylogenetics, contingency, developmental constraints, altruism, kin selection and reciprocal altruism.

effect that use and lack of use has on various parts of our body: but we do not expect assiduous use of the weights to translate into babies with extra muscle. We do not expect this because of a fundamental twentieth century biological belief, known as Weissman's Central Dogma. Simply, this states that information can flow from genes to bodies, but not the other way around: when genes change, so do bodies; when bodies change, genes stay the same.[19]

The idea that evolution might happen through use and disuse, was an idea especially associated in the nineteenth century with Lamarck, a naturalist working some 50 years before Darwin. Though Darwin described Lamarck publicly as 'justly celebrated', in private he was more critical. Still, as already mentioned, in successive editions of *The Origin of Species*, Darwin continued to allow the possibility that natural selection was augmented by other processes. The debate over the scope and ubiquity of natural selection remains a live one today, 120 years after Darwin's death.

Lamarckian inheritance, or the inheritance of acquired characteristics, is a concept much derided in biological circles. The principle's contradiction of Weissman's important thought has led to Lamarck's importance being somewhat forgotten, especially in the UK and the USA, but as any French scientist will tell you, Lamarck was a steadfast evolutionist at a time when the fixity of species was taken for granted. Perhaps of more relevance is the interest today in a set of ideas sometimes called 'neo-Lamarkism', but more often riding under the banner epigenetics, no doubt because of the sour criticism directed at Lamarck by our standard textbooks. The emphasis is not on use or misuse, but on a modern reworking of Lamarck's ideas: that inheritance can happen through processes supplementary to the basic mechanism of transmission of genes.[20] Lamarck of course had no more idea about genes than did Darwin. In epigenetics the emphasis is on the study of cellular processes that control the action of genes. The role of genes in inheritance and development is not

[19] The germplasm view of heredity dates back to August Weissman who in the 1890s, argued that information flows from genotype to phenotype, but never the reverse. In 1958, Francis Crick enthroned this idea as the Central Dogma of molecular biology: DNA makes RNA makes proteins.
[20] When Lamarck is laughed at today, the focus of attention tends to be the giraffe reaching up to the trees, stretching his neck, and passing this acquired feature to the next generation. As far as I am aware, this was never one of Lamarck's arguments. A more accurate account of Lamarck's views was that organisms evolved by 'perceiving' which structural changes would help them to survive the environment. This 'perception' was duly translated into a real change, with a corresponding alteration made in the system of inheritance. The idea sounds strange to us, but Lamark was a pioneer in seeing the importance of environment, and in keeping God out of evolution.

denied, but the precise aim of epigenetics is to research how events at a cellular level – involving proteins, cellular membranes, and hormones, for example – though ultimately dependent for their existence on genes, cannot be precisely predicted from a knowledge of genes. Epigenetics, therefore, is non-reductionist in the sense that it examines causal pathways that it claims are as likely to influence genes, as to be directly influenced by them. Cytoplasmic inheritance is an example of this. When an egg is fertilised by a sperm, the zygote is made up of maternal cytoplasm, as well as maternal and paternal DNA. This maternal cytoplasm has effectively been inherited: it has passed to the next generation. The cytoplasm, by virtue of certain stored mRNA and protein molecules, is now known to have a controlling role in early development, and may have effects lasting into adulthood.

Another example of epigenesis is cortical inheritance. This is seen in ciliated protoctists such as *Paramecium*. These ciliates have rows of cilia, all beating in a co-ordinated fashion in the same direction. Each cilium is embedded in a basal body, in the outer part of the cytoplasm (the cortex) When a piece of cortex is removed, turned and replaced, the cilia carry on beating, but now they beat against the general current. Remarkably, when the *Paramecium* reproduces, subsequent generations inherit this reversed pattern. Yet there has been no change in the nuclear DNA, and it is known that the basal bodies of the cilia contain no DNA themselves.

The significance of epigenesis for evolutionary theory lies in the way that it undermines the primacy of gene mutation as a source of change. For not only may genes be changed by epigenetic factors (in effect it makes sense to talk of 'the phenotype of a gene'); those factors – and the changes they bring – can be inherited from one generation to another. Epigenesis, however, is only undermining of Darwinian theory if one chooses to see that theory as demanding a totally gene-centred world view.

Neo-Darwinism includes the understanding that during evolution, gene frequencies change in a population from one generation to the next. However, not all changes in gene frequency are caused by natural selection, as is shown by the principle of random drift, which is held to be particularly important in small populations. When a population is small, chance events in breeding become important. For example a particularly well-adapted organism, carrying a particularly favourable gene, may simply not come across a mate, or may suffer an untimely death. Its favourable genes, therefore, fall outside the scrutiny of natural selection. Similarly, members of the population that are less well adapted, by happening to come across each other at a lonely waterhole, and by chance avoiding predators and attack by parasites, reproduce and so pass on their genes. Their genes increase in frequency, but by chance (hence the term 'random drift'), not by natural selection: evolution has happened, but not through selection. Random drift is considered a factor in the evolution of small populations. In large populations, the kind of chance events described above still happen, but are statistically insignificant.

An interesting consequence of natural selection is that it produces a genetic load. In a population with variation, some organisms are better adapted than others: they are not all 'optimal'. Natural selection will act against the less-optimal representatives, with the least-adapted organisms being most fiercely selected against, but if selection is very strong, and acting on a population where there is a lot of variation, then there will be a heavy cull of the many individuals that happen to possess the less-than-optimal characteristics. So heavy a pressure on a population might drive it to extinction, suggesting that there must be a limit to the suboptimality that a population can endure. J.B.S. Haldane, the 1930s biologist who worked on the concept of genetic load, thought that the upper limit for gene substitution (a cause of variation) was one substitution per 300 generations.

In the 1960s it became possible to measure variation in blood proteins. For example in 1968 Motoo Kimura compared the amino acid sequences of haemoglobin and cytochrome c in several mammalian species.[21] He found that the amount of substitution was far greater than allowed by Haldane's principle of genetic load. Kimura, therefore, suggested in his neutral theory that a large amount of molecular evolution has no consequence in terms of natural selection. He claimed that a significant number of the gene mutations are neutral. The variation amongst blood proteins is high because a proportion of the variants are neither beneficial, nor harmful. Neutral mutations survive in the population, without being selected for, or selected against: they are 'invisible' to natural selection.[22]

The argument about neutralism has always been over its significance: does it have an important role in evolution? Increasingly, this has been doubted. Haldane's original thinking has been questioned, so that it now appears that high variation is not incompatible with a viable population enduring natural selection. Moreover, in the late 1980s, with new

---

[21] Evolutionary rate at the molecular level, *Nature*, **217** (1968), 624–6.
[22] *The Neutral Theory of Molecular Evolution* (Cambridge: Cambridge University Press, 1983).

technology, it became possible to compare DNA sequences, measuring for example the level of synonymous sites. These are nucleotide changes at the third codon position that have no impact on the meaning of that codon. There is evidence that in *Drosophila*, where the phenomenon has been studied, this kind of mutation undergoes weak selection, even though there are no amino acid substitutions and, therefore, no change in the resulting protein. If selection, however weak, operates at the level of codons, then the neutral theory does not represent nature.

Neutral theorists have, therefore, adapted their ideas in the late 1990s, and moved to the 'nearly neutral theory'. According to these ideas, there is a significant amount of variation that is not neutral, but is only slightly deleterious. Because selection is weak, random drift becomes more important than natural selection in the career of such slightly deleterious mutations. Critics argue that with this modification come other problems, the main one being testability. With the neutral theory, its strong claims made it an ideal null-hypothesis: a mutation was neutral, or it was not. The problem with the nearly neutral theory is not that it is false but that it is difficult to determine empirically the prevalence of slightly deleterious mutations.

Neutral theory has had an important role in the development of molecular phylogenetics. This field is significant because it brings together two aspects of evolutionary theory once largely distinct. One aspect is the study of the evolutionary history of various organisms; the other is the understanding of the mechanism of evolution. Roughly speaking, until the end of the 1960s, the history of evolution was studied by palaeontologists, and its mechanism was left to the population geneticists. That distinction has now begun to be eroded, especially through the techniques of gene sequencing.

DNA sequences, or genes, have evolved, and the closer the similarity in the genomes of two individuals, the closer their evolutionary relationship. Molecular phylogenetics exploits the ease with which DNA can be sequenced to make family trees revealing genetic relationships. A significant difference between gene trees and conventional evolutionary phylogenies is that the former can relate different individuals within a species, rather than being restricted to interspecific comparisons. This allows molecular phylogenetics to track with fine resolution genetic changes over time. In an epidemic, for example, mutations occur in the genomes of viruses, producing changes in phenotypic characteristics such as infectiousness. With a knowledge of the dynamics of these gene changes you can chart the progress of a pathogen through a host population. Such knowledge can then be used for making predictions about the future course of a disease.

In terms of the themes of this chapter, molecular phylogeny raises several important issues. Most obviously perhaps, it shows the way that evolutionary studies can make use of highly reductionist data (gene sequences). Conversely, the field also shows why such data is not going to replace more conventional information from palaeontology. This is shown by the simple fact that molecular phylogenetics, to date, has constructed gene trees only from live, or recently alive individuals. Moreover, just like palaeontology, this is a field that struggles to insert a sense of time into its gene trees. This is where neutralism comes in. The more similar the DNA sequences taken from a sample of the same population, the closer their relationship, but looking backwards, at what point in the past could the sequences be said to 'coalesce' into the one genotype? That moment has to be inferred from a knowledge of the rate of nucleotide substitution (mutation). This is the molecular clock, where mutations tick away as regularly as a well-adjusted watch, accumulating at a steady rate; but for the molecular clock to work as a constant measure, those nucleotide substitutions must be assumed to be neutral in terms of selective pressure.

Evolutionists agree that molecular evolution is modelled well by neutral theory.[23] Equally, the adaptiveness of the phenotypes that we see around us shout out that natural selection is important in nature. Resolving this mismatch is an interesting task for workers in the field, and it is very likely that we will see both reductionist and non-reductionist accounts of evolution collaborating in a final synthesis. Writing in *New Uses for New Phylogenies* Mike Benton compared traditional palaeontology (i.e. morphological cladistic analysis) with molecular techniques: 'it is best to accept the validity of both methods and to enrich phylogenetic studies by informed comparisons between the two'.

Contingency ('chance') is another term that evolutionists use when debating the sigificance of natural selection. Once again, the point at issue is whether an evolutionary change is to be explained by natural selection or by random effects. An important example for palaeontologists has been the mass extinctions that punctuate the history of life. The most famous (though not the largest) is the event 65 million years ago – the

[23] Two texts in particular give a good insight into the problems and possibilities of molecular phylogeny. Masatoshi Nei's *Molecular Evolutionary Genetics* (New York: Columbia University Press, 1987) has a well-written introduction that gives some sense of the biological significance of the field. *New Uses for New Phylogenies*, edited by Paul Harvey *et al.* (Oxford: Oxford University Press, 1996) is a collection of papers that covers a wide range of applications.

Cretaceous–Tertiary or K–T event – that saw the removal of a large proportion of the world's species, including 50% of the marine invertebrates and, of course, the dinosaurs. This was also the time that the mammals, stepping out from under the feet of the dinosaurs, began to flourish. It cannot be denied that extinction is an evolutionary process, nor that it can be caused by natural selection. Yet is extinction necessarily a result of natural selection? The extinction event 65 million years ago has recently been put down to a series of global dramas, including global warming and change in the sea level, and to one cataclysm arriving from space, an asteroid up to 10 km wide hitting the earth on the Yucatan Peninsula in Mexico, which had huge effects on the ecosphere. According to Stephen Jay Gould (1941–2002), a principle advocate of the idea that evolutionary explanations must include contingency (or chaos), it is simply incredible that the massive weeding out accompanying these catastrophes can be put down to natural selection alone.[24] The argument can be simply put: if a few billion tonnes of dust are put into the atmosphere and the oceans' temperature shifts by 10 °C, the changes are not going to select between the individuals of a finely varying population, but will instead strike brutally and indiscriminately. Chance explains why, quite suddenly, after 150 million years, all the dinosaurs were extinct, not simply the maladapted few. It also explains why the mammals were able to initiate their prodigious adaptive radiation, a process that has led to the primates, and to human beings able to ponder the mechanisms of evolution.

As with my previous examples, the theory of contingency can be attacked on the grounds of lack of evidence. Contingency seems to fit the facts, but as so often with evolution the facts themselves are questioned. For example while the extinction event of 65 million years ago is popularly conceived as a brief period where serious havoc quite suddenly caused intense difficulty, there is evidence that the event was, though brief in relation to geological epochs, not actually an instant occurrence. For example, the ammonites – a group of marine cephalopods – went extinct in the K–T event, but had been declining for some 15 million years. In this case, perhaps, the asteroid from space was not a prime cause of their final extinction, but merely the *coup de grâce*. Up to that time, one can imagine, natural selection was operative, was discriminating between ammonites, and driving some to extinction, before the survivors were overwhelmed when the asteroid struck.

---

[24] *Wonderful Life: the Burgess Shale and the Nature of History* (New York: W.W. Norton, 1989).

It is often pointed out that developmental constraints will limit the powers of natural selection. The study of form, morphology, used to be an important field of biology, especially in Darwin's time, but lost ground to genetics and the more reductionist styles of biology in the 1920s and 1930s. However, thanks to modern, genetically inspired advances, an important insight can be drawn from developmental biology, and in particular from the genes that are known to control development. It seems that these genes are highly conserved. For example the genes that control the top–bottom orientation of a developing *Drosophila* larva are also at work in a human fetus. These are the so-called homeotic genes that work together to produce fundamental aspects of development. No doubt these genes have themselves been selected for, but we can view them also as constraining the amount of variation of form that could ever appear. Homeotic genes are at work in one of the best-known patterns in nature: the tetrapod limb. The fact that a horse limb, a whale limb and a human limb share many similarities is usually explained by the Darwinian theory of common ancestry: they all evolved from a common ancestor, adapting as time went by to the particular environment. There is, however, another way of looking at these limbs. There is experimental evidence that the basic form of a growing limb can be 'jumped' from one type to another by varying certain chemical gradients. In other words, it is not the case that it is genes that determine form: they are working in combination with, perhaps even being controlled by, chemical events occurring at cellular (or higher levels). Work of this sort sometimes discusses 'morphogenetic fields', a vague term for the overarching chemical and cellular events that could determine the behaviour of genes. The implications for the Darwinian explanation of evolution are clear. The variety that natural selection works on is not determined simply by the variation thrown up by mutation and recombination: there are important processes at work in cells, and in tissues and organs, setting constraints on how that trait develops. The development pattern, in short, cannot be explained simply as a result of natural selection.

The concept of developmental constraint was vividly conveyed in the celebrated essay 'The spandrels of San Marco', written in the late 1970s by the palaeontologist Stephen Jay Gould and the geneticist Richard Lewontin.[25] They saw in the roof of the Venice basilica of San Marco a

---

[25] 'The spandrels of San Marco and the Panglossian paradigm: a critique of the adaptionism programme', *Proceedings of the Royal Society of London Series B-Biological Sciences*, **205**, (1979), 581–98.

feature that could usefully support the idea that not all that is useful in organisms arose by natural selection. When fan vaulting supports and fills a normal rectangular ceiling, triangular spaces remain. The spaces are called 'spandrels', and are used to great effect by cathedral builders, who fill them with mosaics, bosses and other decorations. Gould and Lewontin show that these spandrels are a by-product of the main architectural design, but have been put to good use anyway. The existence of the spandrels is not explained by the decorations. As an example from nature, Gould has pointed to snails, which build their shells by coiling them round an axis. This leaves a cylindrical space, called the umbilicus. A very few species of snail use the umbilicus as a brooding chamber for eggs: the umbilicus has become a 'secondary adaptation'. Yet according to Gould, the umbilicus cannot be considered as resulting from natural selection, because it has arisen passively as an accident of shell coiling.

So far I have described natural selection as acting on the variation within a population. Individuals, which, by virtue of the chance variations they have inherited, are better adapted, are statistically likely to leave a greater number of offspring. Those favourable characteristics will, therefore, be better represented in the next generation. Described like this, we see that natural selection is highly individualistic: the highest-reproducing individuals are, by definition, the most successful.

Ever since Darwin, commentators have seen something appalling in this feature of natural selection, for it seems to allow no room for altruism, defined biologically as behaviour that decreases the reproductive success of one organism (the altruist) in favour of another. Making the reasonable assumption that most animal behaviour has a genetic base, altruism can never evolve by natural selection, because any gene that produces altruistic behaviour (or indeed any altruistic animal) is by definition going to be less successful. As discussion of animal behaviour tends to provoke ideas about human behaviour, there is something disturbing in the belief that natural selection, nature's prime motor of evolution, is austerely and inherently individualistic. We see plenty of selfless behaviour in humans, and rarely imagine it conflicts with biology. We see plenty of co-operation in nature too: lions hunt in groups for example. Altruism also exists. Worker bees, unable themselves to reproduce, slave away on behalf of the queen's own offspring. The bird acting as sentry for its flock calls out to its companions that a hawk is on the horizon; it saves them, but perhaps calls danger down on itself. The question is raised: how can natural selection allow altruism to evolve?

Examples of altruism in nature led some naturalists in the 1960s to invoke the concept 'group selection'. Here, a trait is selected because it benefits the group rather than the individual. A group of altruists might indeed last longer than a group of selfish individuals, and by emigrating, they might produce descendent subgroups, made up of individuals carrying altruistic genes. An example favoured by V.C. Wynne-Edwards, an early group selectionist, was that of a group of animals avoiding extinction by showing reproductive restraint when the local food supply becomes scanty.[26] Such behaviour, according to Wynne-Edwards, would make the group more likely to survive, and so would classify as an adaptation that benefits the group. This and other examples drew a sharp reply from G.C. Williams, whose 1966 book, *Adaptation and Natural Selection*,[27] became the inspiration for Richard Dawkins's extremely influential *The Selfish Gene*.[28] Williams argued that there is a danger for a population of altruists. While it may be true that such a population is less likely to become extinct than a selfish one, consider what would happen as soon as a selfish individual invades the altruist group. The selfish individual, congenitally unable to defer to the needs of others, will be more successful, leave more offspring, and its inherited selfish behaviour will overwhelm the population. We see then that the argument between individual selection and group selection is one over the biological roots of altruism. For altruism exists in nature, and it does so in organisms that have no moral sense: ants for instance. The question is forced on us once more: how can natural selection produce altruism? Two compelling elaborations of the theory of natural selection are thought to solve the problem largely: kin selection and reciprocal altruism.

If the animals within a group are related (they are 'kin'), then altruism can evolve by natural selection. For if we focus on the survival of genes, as well as on that of individuals, then genes that produce altruistic behaviour are likely to benefit (be better represented in the next generation), so long as the altruistic behaviour is directed at relatives. Siblings, for example, have a 50% chance of sharing any gene, exactly the same proportion as found between parent and offspring. Helping your brother to survive helps you to survive, or at least, it helps your genes to survive. In this case, the group does benefit, as well as the individuals within it, but as a result

---

[26] *Animal Dispersion in Relation to Social Behaviour* (Edinburgh: Oliver & Boyd, 1962).
[27] Princeton, N.J.: Princeton University Press.
[28] *The Selfish Gene* (Oxford: Oxford University Press, 1st edn, 1976).

of individual selection, rather than as an example of group selection, in which it fails.

Organisms help each other, even when they are not related. If we accept that altruism cannot evolve except by natural selection, under what circumstances could altruism towards non-relatives emerge? In the theory of reciprocal altruism, such behaviour may directly benefit individuals (and their genes) if it occurs on the basis of one good turn deserves another. Robert Trivers developed this theory.[29] In one of his examples he discussed the fairly common symbiosis found where one fish (the cleaner) removes parasites from another, usually larger fish (the host). Clearly both benefit, yet there is altruism on the part of the host. It does not gobble up the cleaner after being cleaned (which would be easy, as cleaners go right inside the mouth and gill cavity), and so foregoes the apparently sensible option of having a good clean, followed by a good meal. Moreover, there are many examples of hosts going to some pain to warn the cleaner that it is about to move off, and of hosts helping a cleaner by chasing off predators that pose a threat. Yet though the host is being altruistic in terms of some behaviours, its overall strategy is beneficial to its reproductive success. There is evidence that hosts use the same cleaner, returning to the same cleaning station again and again.

Kin selection and reciprocal altruism are models that 'attempt to take the altruism out of altruism'. The behaviours that led Wynne-Edwards to claim the existence of behaviours that existed for 'the good of the species' have, it seems, been explained as instances of individualistic natural selection. Today, few biologists work on the concept of group selection, and those who do are considered somewhat heretical. Yet it is widely agreed that in some circumstances group selection may be possible, for example if selfish individuals can be destroyed, or prevented from reproducing. If altruism is to evolve by group selection, what is required is that groups whose individuals are altruistic protect themselves from infiltration by, and are separated from, groups made up of selfish individuals. With groups having a relatively long 'reproductive period' compared with individuals, some active management will be necessary to prevent mutant 'cheats' – carpetbaggers – from taking over when they arise. The reason why group selection is considered to be a rare event in nature is that groups reproduce more slowly than their member individuals. A mutant

---

[29] *Social Evolution* (Menlo Park, Calif.: The Benjamin/Cummings Publishing Company, 1985).

selfish individual will spread its genes through an altruistic group long before that group has had the chance to generate an altruistic descendent of its own.

3.10 The 'unit' of natural selection

We have seen that there is debate over the relative significance of individual, kin and group selection. A final question remains. The orthodox beliefs described in the previous section see adaptations as favouring individuals, but in ways diverse enough to produce in some organisms, altruism, or more correctly, the illusion of altruism. In those discussions we frequently referred to genes as being the significant entity, because the same genes can be shared by many individuals. The question thus arises: on what does natural selection act? Clearly, it is individuals that die and reproduce, and manifest the kind of complex relations that are both individualistic and altruistic at the same time. Yet when we come to model natural selection mathematically, shall we consider genes as the entity selected, or individuals, or even groups? Again, there is an orthodox opinion on this point. Evolution by natural selection is considered to lead to a change in gene frequency. In one of many disputes between evolutionists, Stephen Jay Gould has affirmed that genes do change in frequency owing to natural selection, but that this is a matter of 'bookkeeping'. When an adaptation gains ground within a population, this is because of real life and death weeding out one organism but not the next. Other biologists, including Richard Dawkins[28] and John Maynard Smith,[30] have claimed this is not a helpful way to look at the issue. For them, natural selection can only differentiate between entities that are permanent, and that means genes. The point is that genes are permanent, while individuals are not. Only gene frequency changes can be considered to be evolution, because only genes have the requisite permanence. Organisms are important because they inherit the genes and pass them on; but they are not themselves the site of action of natural selection.

This is the gene-centred approach to biology. In the discussion presented here the value of seeing genes as the unit of selection might be that it makes mathematical modelling possible. Yet whatever the merits of that approach, we see now in biology another type of gene-centredness: the belief that of all the causes we might investigate, genetic causes are the most important. For example the field of evolutionary psychology argues that

[30] *Evolutionary Genetics* (New York: Oxford University Press, 1989).

our own behaviour has evolved by natural selection and can, therefore, be examined as a series of adaptations. Sexism, violence, rape and murder are just some of the features of life that are, from time to time, given an evolutionary explanation. As we shall see in the next chapter, such explanations are invariably controversial. The idea that our behaviour is as much evolved as cultural raises many questions, some of which concern the way in which we see our relationship with other organisms. Are we like them, or are we alone? An exploration of that relationship forms the substance of the next chapter.

3.11 The Darwin wars

This chapter has considered Darwinism, and certain ideas that supplement it. I have not made any attempt to evaluate the current status of Darwin's work, nor how new ideas are influencing it. Debates about Darwin's legacy, however, are remarkably fierce, and have been described as the 'Darwin wars'. Though there are many biologists who have pitched into these battles, two names stand out: Richard Dawkins, famous for his early book, *The Selfish Gene*,[28] and Stephen Jay Gould, also best known for his writings. Many books, many column inches in magazines, and many seminars, have taken positions around these two figures. It would be extremely difficult and time consuming to represent fairly their dogged arguments and their frequent insults. In essence, Gould's work, on the one hand, represents the point of view that Darwinism is not of universal validity in biology, let alone in the social sciences. Dawkins, on the other hand, sees Darwinism as a kind of general theory of everything, applicable to all of biology, and also to plenty of human behaviour. In addition, Gould tends to downplay the centrality of genes as a focus for natural selection, while Dawkins takes a more gene-centred view.

Put like that (and this is a simplified version of their views) there does not seem an obvious reason for the fierce battles that have broken out, and these have been fiery even by academic standards. Perhaps the clearest focus for the dispute concerns evolutionary psychology, discussed more completely in this book at the end of Chapter 7. For evolutionary psychology takes the view that many human behavioural traits have evolved by Darwinian mechanisms, that is, by natural selection. Specifically, when the human mind was evolving, selection pressures acted on any inherited traits concerning aggression, sexual behaviour, aesthetics, diet, ability to enjoy play, and so on. Thus (once again, to simplify slightly) evolutionary psychologists regularly pronounce on rape, or modern art, attempting to provide a 'Darwinian interpretation'. Part of that interpretation, of course, is that any behaviour that we have 'inherited' is part of our nature and, therefore, may be not so easy to change. The forces who congregate around Gould consider this simple nonsense, for a number of reasons. One is that we have no idea what man's early environment was like and, therefore, cannot assume what the selective pressures might be. Another is that the link between genes and traits such as aggression, or sexism, is simply unknown, even if one accepts aggression as a trait. Therefore, the inheritance of these genes, and their subsequent meaning, is simply conjecture. Perhaps the main objection, however, and the reason for the passion, is the perceived social agenda of the evolutionary psychologists. According to the Gouldians, they are right-wing pessimists, who downplay the significance that the environment can have: education for children, rehabilitation for offenders, support groups for drug addicts. Needless to say, evolutionary psychologists deny this, and claim to be pragmatists, or realists.

To bring all this into focus, we can consider some research into death rates amongst stepchildren. The two psychologists Martin Daly and Margo Wilson, who would certainly line up alongside Richard Dawkins, investigated homicide rates amongst children.[31] They found that stepchildren are far more likely to be killed than natural children. Moreover, when they are killed, the death tends to be violent and deliberate. When natural children are killed, it is often by neglect. Clearly, there is a Darwinian interpretation of this. Stepfathers are not related to their stepchild, and so, faced with the inevitable irritation of the very young, soon lose control: any screaming child is hard to cope with, but if it is your own the experience will not be so bad. Yet for this to be a truly Darwinian mechanism, one would have to show that behaviour like this happened 30000 years ago, was advantageous to the adult couple, and could be inherited. Needless to say, this evidence is hard to find. But as ever in evolution, the argument will carry on, whatever the state of the evidence. For evolution is not only about the way the world is, but the way we would like it to be.

[31] *The Truth about Cinderella: a Darwinian View of Parental Love* (New Haven, Conn.: Yale University Press).

4
Biology and animals

4.1 Ethics in science: an introduction

This chapter is about ethics. I want to show that studying biology not only makes intellectual demands, but ethical ones too. In practice, an ethical issue in biology arises when people disagree on whether it is right or wrong to do a particular piece of research. The operative words here are 'right' and 'wrong'. We are used to seeing biological research as right or wrong in a quite different sense: right in the sense of true, wrong in the sense of untrue. Here we are using the words differently: we are impling that unethical behaviour in science is wrong, in the same way that we consider that lying, or murder, are wrong. Just as in war we tell lies and kill people, and believe that we are doing right, so in the study of ethics in science, there is a lot of space for argument and disagreement, a disagreement never completely resolved by the facts that the two sides marshal as evidence. This chapter looks at the mix of facts, history and beliefs that constitute ethical argument.

Biologists, like all scientists, like to think that they are making the world a better place. The role that genetics plays in modern medicine might be used as an example of how biological knowledge can increase health. Yet biology and its applications are sometimes viewed not with appreciation, but with scepticism. It is not only the press, politicians or the public who urge caution whenever a new development in genetic engineering, or in fertility technology, is announced: so do many scientists. When Dolly the sheep took all the headlines in the spring of 1997, having been revealed as the first mammal to be cloned from an adult cell, her maker, Dr Ian Wilmut, stated that he could see no reason for using the technology to clone humans. Wilmut was responding to an anxiety, voiced also by President Clinton in the USA, that such biological knowledge could be used improperly. A technology that could produce genetically identical farm animals, might also be used to produce genetically identical humans. They implied that while cloning sheep might be 'right', cloning humans would be 'wrong'; this is clearly a moral claim.

Fears of what science can do are not a recent media invention. The story of Victor Frankenstein's monster was created by Mary Shelley during a Swiss holiday in the summer of 1816. Shelley was inspired by the 'new' chemistry of the day, which had begun to research and organise the elements. These elements were seen by some philosophers (for example, Charles Darwin's grandfather Erasmus Darwin) as the material substance of life, just as they were of tables and chairs. Given the new successes of chemistry, it seemed plausible that the manipulation of elements might lead not only to the manipulation of life, but also to its creation. Shelley's story drew on such ambitions, and also on contemporary ideas about the links between electricity and life. The idea that electricity had something to do with living things was at that time being dramatically revealed to the public by a bizarre scientific fad, in which already executed criminals were electrocuted in a type of circus entertainment. The twitches and shudders of the corpses suggested that life was in some way electrical; such ideas were being explored at the time that Shelley wrote her story. The Frankenstein story is remarkable not only for its literary merits, but also for its persistence as an alarm bell, to be sounded vigorously whenever science is perceived to be 'going too far'. Of the sciences, it is biology, and particularly the fields of development and genetics, which are most likely to get branded with the sign of Frankenstein. In 1999, when British newspapers suddenly gave massive and hostile coverage to genetically modified foods, the slogan 'Frankenstein Foods' (or 'Frankenfoods') was adopted even by the sober Daily Telegraph; no one needed a lesson in nineteenth century English literature in order to catch the implication.[1]

I mention Frankenstein to show how a discussion of the cost and benefits of biology can veer into areas usually considered to be 'outside' science. For example the generally suspicious attitude of the British media to genetically modified (GM) foods was probably responding as much to ideas about such food being 'unnatural', and to the only-too-obvious link

---

[1] The links between Shelley's story and contemporary science are best covered in Jon Turney's *Frankenstein's Footsteps* (New Haven, Conn.: Yale University Press, 1998). This account surveys the sensitivity of the popular consciousness to the attempts of scientists to manipulate life, or to create it. Thus, though scientists moan at what they see as crude headlines in the press, it does not seem as though the media is itself responsible for public apprehensiveness of the powers of scientists.

between GM foods and big business, as to considerations about the ambitions of science. Though the newspapers dutifully published details of the actual scientific doubts that could be raised, such as the question of genes moving into wild pastures, and creating 'super-weeds', it was the non-scientific aspects of the issue that prompted the banner headlines. Here was an opportunity to give a large multinational company a bloody nose: how dare they spike our British food, the editorials admonished. There was altogether less interest in the facts of cross-pollination, or in what 'genetic modification' actually means. Many scientists view this phenomenon of selective reader interest as a damaging and even dangerous distortion, evidence that the public knows too little science, and that the mass media run by arts graduates is hostile or simply indifferent to the real work of science.

There is evidence that it is quite common for scientists to view with suspicion society's interest in their work. A set of guidelines for science journalists, published by the Royal Society and the Royal Institution of Great Britain,[2] might fairly be interpreted as assuming that public suspicion of science occurs simply because journalists fail to explain scientific insights clearly. The idea that a non-expert could have a legitimate reason for disagreeing with a scientist, and that investigative journalism might have a role in challenging scientists, is strongly downplayed. In a similar vein, the Director of the Royal Institution of Great Britain, Baroness Susan Greenfield, has declared:

> Since the media have as a priority a good story, and since scientific discovery is a gradual process of refinement of falsifiable hypotheses, it is almost inevitable that only the quirky and sensationalist science story gets aired . . . The result is that the public are unduly panicked, insufficiently informed and view scientists as dysfunctional nerds or power-crazed robotic megalomaniacs with no normal emotions.[3]

Yet if biology can affect the well-being of the world, then a public curiosity, however non-technical, is guaranteed. Biologists – and that includes students of biology – must be prepared to account for their work, and discuss it in a variety of arenas, including television and radio studios, newspaper columns, debates and talks. Such science communication, like all communication, is a two-way thing: it is much less successful when scientists

[2] http://www.royalsoc.ac.uk/news/guidelines.htm.
[3] The quote is taken from the Engineering and Social Science Research Council (ESRC) report on science and the media *Who's Misunderstanding Who?* (http://www.esrc.ac.uk/esrccontent/connect/indexpub.asp).

simply tell the public what they think they should know, far more successful when the scientists actually find out what the public know already,
what more they would like to know, and what they do not like. Moreover,
there is not one 'public', but many (for example students, patient groups,
children, farmers, trades unionists). Each 'public' has its own particular
knowledge and interests, so good science communication is likely to be
varied in style, but always with the aim of maintaining a dialogue; remember that the words 'communicate' and 'community' have the same root
Latin word *communicare* (to share).

The easy goading of modern science by the Frankenstein story reveals
as much about some popular conceptions of nature, or of change, as of
science. To some extent then, the attitude of scientists to banner headlines
(dismissed as sensational and inaccurate) is understandable. However,
there is another common feature in ethical debate in science. This is the
claim that whatever a scientist develops in the laboratory, its eventual use
is the responsibility of governments, hospitals, industries, armies or other
customers of the products of science, and not that of the scientist. In the
middle of the twentieth century, during the Second World War and then
the Cold War, it was the physicists who were closest to this problem. Wars
are fought with bombs and bullets, radar and surveillance systems. There
is a great deal of physics in a war, but it is politicians and generals, not
physicists and technicians, who make the military decisions. Yet after the
Second World War, and especially after the nuclear bombing of Hiroshima
and Nagasaki with fission devices, the fact that physics research could
have so important a military consequence led to an increased debate about
the social responsibilities of science. Scientists were necessary for the development of a nuclear capability in the USA in first place: Albert Einstein
wrote to President Roosevelt in 1939 warning him of the potential influence of nuclear physics. Los Alamos, the huge nuclear research facility in
New Mexico, was set up to develop and build atomic bombs. The world's
first detonation of a nuclear device, on 16 July 1945, in a test firing in
New Mexico, was supervised and analysed by scores of scientists, some of
whom, after the war, were to campaign against the nuclear arms race between the USA and the then superpower, the Soviet Union. The most famous of these morally concerned physicists was Robert Oppenheimer, the
first Director of the Los Alamos laboratory, and a national hero. Soon after
the war he found himself defined as a security risk by US intelligence –
simply because he opposed the development of the far more powerful
next generation of nuclear devices: the thermonuclear or fusion devices.

In the middle of the twentieth century, the ethical problems of science seemed to come from the physical sciences. A fair chunk of the second half of the twentieth century was to be spent obsessed with the nuclear arms race, with the Cold War, and with the possibility of nuclear war or MAD (mutually assured destruction). When people worried about the powers of science and scientists, and wanted to draw up a balance list of the benefits and hazards of science, the mountain of worries about physics loomed large. There was a standard cliché: science is a double-edged sword. Everyone agreed that scientists were not responsible for the atrocities of war, but there was argument over whether it is right that scientists should see themselves as insulated from the applications of their research, and the argument continues to this day. The 'it's nothing to do with us' view of science sees it as an exercise of pure research, whose discoveries are unpredictable, and which depends for its success on a proper isolation of scientists from the pressures of commerce or government. According to this view, the insights streaming out of the universities and research centres are simply 'knowledge'. What is done with that knowledge, and any ethical responsibilities attached to the application of that knowledge, is the responsibility of those who live in the real world: politicians, pharmaceutical companies and arms manufacturers. This is the idea of the scientist as undisturbed, isolated from ethics, who may be seen either as 'the mad scientist', or thought of as an embodiment of 'academic freedom'. It is also the idea behind the split between science and technology, or between pure science and applied science. According to the 'it's nothing to do with me' view, science, unlike technology, is not troubled by moral conflict.

I have sketched out these famous pitfalls of ethical discussion of the twentieth century, but do not intend to analyse them further. A very full exploration would be needed to do justice to the problem of judging a solid state physicist's responsibility for modern warfare. Yet it is worth noting that if physics was the ascendant science of the mid twentieth century, the start of the twenty-first century sees biology as the science of greatest impact. This pre-eminence is due to genetics, and to the ability of biologists to manipulate genes and their expression in cells. The ethical debates that surround science are now influenced by biology rather than by physics: what do we think of xenotransplantation, should genetic information be confidential, can society insist that a person has a genetic test? It may be that, after the attacks on New York on 11 September 2001, biology will replace physics as the main field for 'weapons scientists'.

For a biologist, there is one ethical issue that cannot be avoided. It is the question of animals. Many biologists intervene in the lives of animals. Whatever the eventual benefits of these interventions, the biologist who works with animals has to reflect, to a greater or lesser extent, on the fact that a great deal of biological and medical understanding has come about by killing organisms. I am of course talking about vivisection, and an analysis of this forms the rest of this chapter.

4.2 Interventions: an introduction

To be able to think about morals, you have to feel that you have an ability to make a decision, that you can think through a problem, and then commit yourself to some action as morally right. If you feel that the study of biology involves making moral decisions, then you are likely to believe also that becoming good at biology involves more than learning facts and picking up research techniques Yet what is your main impression when you survey a university life science course outline, or a biology textbook? Are you excited at what lies ahead, or do you feel daunted by the sheer volume of detail? Whatever you come to feel about your studies, there is no question that the absorption of a fairly large amount of factual material is going to feature strongly. Learning a great amount of such information is part of becoming a trained biologist. The problem comes if you begin to feel that it is your memory that gets you through a biology course, rather than your intellect. Overuse of the memory leads to the view that biology – and biologists – simply collate data about nature, like scribes copying a manuscript more or less faithfully. We saw in Chapter 1 that biological research is not simply a matter of opening your eyes and seeing what is before you. Active intervention is required, with devices and with theories. These actions might not affect any living thing, for example when you count the mussels on a beach affected by oil pollution. They can also be quite drastic, as is shown by my first example, research into ecological succession.

Succession is the term given to the way an area of land or habitat changes over time in its variety of species. The phenomenon has been closely analysed and has generated a terminology. Primary succession is the term given to the changes occurring in an area that has been lifeless: a retreating glacier or a newly formed volcanic island. Secondary succession takes place after a disturbance where the soil has nonetheless been left intact; it is understanding this type of succession that provides such a good example of the power of the biologist to protect nature. For it is secondary succession that interests the ecologists and the politicians and the inhabitants of Amazonia, who work to re-establish the vegetation in a sector of tropical rainforest laid waste by fire and by farming. Succession is characterised in its early stages by opportunistic 'fugitive' species that have high fecundity but also high mortality and a short life (the r-selected species). These spread quickly, but fade as a wider variety of less fecund and longer-lasting species, (the K-selected species) move in. As the species thrive or fade, they themselves have an effect on the environment, perhaps by stabilising the pH of the soil. A change like this is called facilitation, because it allows other species to gain a hold and flourish. After a sufficient period, some kind of stability or equilibrium is obtained, called a climax community. In the temperate regions of Europe and North America the climax community will be forest, and land will tend towards this state. According to some biologists, a climax community is not necessarily best understood as in stable equilibrium. Unpredictable events such as land disturbances or fire, or chance variations in the dispersal power of species, may lead to constant change or cycling within the area, the so-called polyclimax.

The phenomenon of succession, described like this, sounds like an established body of knowledge. These are the terms you may have to memorise. Yet this small sample of biological understanding required experiment. Such research is not simply a matter of watching habitats, much less a matter of reading textbooks: action is required, and sometimes it is significant. A famous example of this took place in 1964, in the middle of the Cold War. Wanting to know how atomic explosions would affect the biosphere, the US Government had taken to irradiating selected habitats throughout the USA. Under the leadership of the ecologist Howard Odum, a portion of tropical rainforest in Puerto Rico was subjected to this treatment. The irradiated hectares lost many of their species, and the resultant succession was tracked. The richness of the habitat before it was destroyed, and its subsequent recovery, gave information about the species involved, and their eventual distribution. The point is that radical intervention was needed before succession could be described. Similar experiments on islets of the Florida Keys in 1965 involved fumigating small islands with methyl bromide. This general biocide kills all the arthropods, but degrades quickly, allowing the organisms back; the re-colonisation can then be studied. In such experiments, species numbers on the islands return after a year to their pre-experiment levels. The type of species varied slightly, perhaps reflecting chance events in terms of which arthropods happened to arrive on the island.

This research produces knowledge useful to conservationists, but reminds us that biology is better described as an experimental science than a descriptive one. To investigate the lives of animals and plants, you are likely to change those lives. Moreover, biologists may decide that animal suffering is inevitable in some experiments. This is a moral decision, in the sense that the rightness of vivisection is said to lie in its benefical influence on human health, and on biological knowledge. Vivisection, in other words, is right because it produces something morally good. Not everyone agrees with this viewpoint, and I shall now examine these arguments in more detail.

4.3 Vivisection and dissection

Definitions

The word 'vivisect' comes from the Latin *vivus* (alive) and *secare* (to cut). However, vivisection properly understood is not simply the cutting into live animals. In the UK and the USA (as in most other countries) it is a legal practice, regulated by law, where experiments are performed on living animals for the purpose of science. Vivisection is a different thing, therefore, from the slaughter of farm animals, the treatment of animals by vets, or the cruel abuse of pets by their owners. Included as examples of vivisection are experiments where cutting or surgery is not involved: the use of microbiological or chemical agents on animals; the manipulation of diet; or psychological testing. 'Dissection' is a word that comes from the study of anatomy; it means the cutting of animals, including people, so as to study or gain access to, organs and tissues. Thus a surgeon dissects in order to reach a point of surgical interest, as does a student who extracts the gut contents from a killed locust; but however skilful the work of the butcher preparing Mr Jones' cut of beef, his work is not dissection.

It is important to be aware in all of the following that the zoological definition of animal is much more inclusive than that used by the vivisection laws and by most animal rights campaigners. According to the US Animal Welfare Act: 'The term animal means any live or dead dog, cat, monkey (non-human primate mammal), guinea pig, hamster, rabbit or such other warm-blooded animal.' In this chapter I shall follow the Act's conventions, rather than the biologists' classification: I shall use the word 'animal' to mean mammal, though we will find that ethical arguments about the treatment of animals require some clarification concerning which areas of the animal kingdom are being fought over.

Justifications

The argument in favour of vivisection is that human health depends on such experiments. What medical knowledge and cures we possess has been vitally dependent on evidence taken from experiments with animals. This is not to say that all our medical knowledge depends on vivisection, but simply that such research is an important tool in biomedical research, and will remain so. Nor is it hard to find evidence for the argument: your very existence might itself be evidence. Smallpox, cholera, whooping cough, diphtheria, polio, tetanus, puerperal sepsis, rheumatic fever and tuberculosis are some of the diseases brought under control by a combination of vaccines and antitoxins. Consider drugs such as insulin and the immunosuppressants, or the techniques of anaesthesia, analgesia, and cardiovascular surgery; vivisection was involved in the development of each of these. Similarly, animal research is basic to research into one of today's major killers: cancer. Cancer research, of course, is very diverse; one aspect, the study of the immune system, not only explores how cancer cells evade the body's defences, but has also led to the laboratory development of monoclonal antibodies, which can be used for targeting specific cells. Monoclonal antibodies were developed by inoculating mice with a particular antigen, taking cells from their spleen, and fusing those cells with cancer cells, so as to produce a hybridoma cell, which can be cultivated indefinitely and used to generate pure antibody. Another tool for cancer research is the transgenic or 'oncomouse': genetic engineering has made it possible to insert a cancer-producing mutant gene into mice, so that the transgenic mouse, in developing cancer, grows human tumours. According to many, the usefulness of transgenic animals in medical research is actually likely to lead to an increase in the number of animal experiments, at least in the short term.

The basic defence of vivisection, therefore, is to link biomedical advances with animal experiments. The biomedical advances include fundamental knowledge acquired centuries ago (the circulation of the blood) and more recent physiological insights (synapse structure and function, for example). As I have already said, potent examples can be taken from the great list of diseases that have been brought under control this century. Perhaps even more persuasive of the need for animal experimentation, if it is accepted that such research leads to better medicine, is to remind ourselves that we start the twenty-first century in the company of many dangerous diseases, for example cancer, AIDS and malaria. However, with science now able to devise alternatives to vivisection, such as computer modelling or tissue culture, are we not able now to dispense with vivisection? No, runs the argument for animal experimentation, for physiological processes studied in isolation, mimicked by computer, or simplified by tissue culture, may behave differently from those in a whole animal, or in a whole human. Indeed, a fundamental assumption of the biomedical animal experiment is that results emerging from the experiment gain credibility by being drawn from a living system whose complexity mirrors that of a human. Similarly, animals (remember that we are defining them as non-human mammals) are judged sufficiently similar to humans, especially as regards their biochemistry, for them to be used reliably as 'human models'. That judgement of course is itself a result of science. Darwinism explains why humans might be similar to other animals, and work in ethology, anatomy, biochemistry and genetics provides many confirming observations.

The justification of vivisection is alive to the fact that it is the suffering of animals that arouses public opposition. The reassuring words put forward about this suggest that in today's highly regulated environment suffering is in fact kept to a minimum. Under the law, and according to guidelines, researchers are expected to make every effort to reduce suffering – by using fewer animals, by using anaesthesia, by looking for alternatives to animal models, by good practice in keeping animals, and by the euthanising of animals likely to suffer unacceptable post-operative suffering.

Criticisms

The argument against vivisection asserts it to be cruel, unreliable and unnecessary. Like the basic justification, it is not hard to find supporting examples, both ancient and modern, to insert into the argument. A good place to start would be Peter Singer's ground-breaking *Animal Liberation* and often described as the 'bible' of the animal rights movement.[4] In Chapter 2 of Singer's book we read of experiments by the US airforce that tested the motor co-ordination of irradiated chimps. Errors by the strapped-in animals were punished by electric shock. Examples like this, drawn from science and from animal husbandry, crowd Peter Singer's book, and make sorrowful reading. Needless to

[4] New York: Avon Books, 1975.

say, as the literature that campaigns against vivisection points out, cruelty to animals is not a feature restricted to modern life. The ancient Romans are an example: an entertaining afternoon in the second century might have consisted of a trip to the Colosseum where the thrill might be to watch elephants tormented to death by tigers. When the Roman emperors converted to Christianity, massacring slaves for fun was banned, because human life was redefined as sacred. The animal contests continued however, fading only when the Empire itself had become disordered enough to make the acquisition of exotic animals next to impossible.

In such accounts of animal cruelty over the centuries, an important figure is the seventeenth-century French philosopher René Descartes (1596–1650) who argued that animals have no awareness of pain, indeed no awareness at all. Though there is some academic debate about whether he actually believed such an extraordinary thing (he was by all accounts friendly to his dog), and though Descartes has hugely influenced modern philosophy, it is customary to condemn his influence on man–animal relations as wholly baleful. His beliefs about animal awareness endured, and must be considered partly responsible for some awful nineteenth-century physiological experiments and, in response, the first sustained anti-vivisection campaign. It was during the nineteenth century that science became professionalised, with university departments, professors and students. Physiology was then a leading subject in the life sciences, much more than now, and vivisection was standard – even before anaesthetics had been developed. The accounts of dogs and other animals dissected alive and without anaesthetic are more chilling than the tales from Rome because we recognise more immediately the rationale behind the experiments: in spite of the barbarity, there is a modern feel, and a modern justification. These are experiments that are important to the history of science, and whose results are still learnt by schoolchildren. The analysis of the reflex arc, for example, a staple of biology examinations for 16 year olds across the world, was one of the triumphs of nineteenth-century physiology, and of vivisection. The great nineteenth-century French scientist Claude Bernard, who we label 'the father of physiology', and recognise as an experimental genius, relied entirely on vivisection, and must have caused great suffering to his animal subjects. He was the first to exploit the drug curare in experiments. Curare, a natural extract from certain tropical hardwoods, paralyses animals. Bernard would inject dogs and cats with curare, which abolishes the motor response, but does not interfere with sensory nerves. Thus the animal feels the pain, but cannot move. Bernard and his assistants used such experiments to demonstrate the crucial division in the vertebrate nervous system between nerves carrying information to, and from, the central nervous system; anatomical dissection on dead animals would not have shown this, but since curare was discovered before the advent of anaesthesia, the animals that brought us this knowledge ended their lives in agony.

The English were the first campaigners, and the first politicians, to inquire systematically into the moral problem of vivisection. In an account to the Royal Commission of 1875, set up by the Prime Minister Disraeli (allegedly on the insistence of Queen Victoria) and tasked with bringing forward legislation, a Mr Hoggan gave an account that described the work of 'a very eminent scientific man in Paris, M Paul Bert'. According to Mr Hoggan, curare was used to

> render a dog helpless and incapable of any movement, even of breathing, which function was performed by a machine blowing through its windpipe. All this time however its intelligence, its sensitiveness, and its will, remained intact; a condition accompanied by the most atrocious suffering that the imagination of man can conceive. In this condition the side of the face, the interior of the belly, and the hip, were dissected out in order to lay bare respectively the sciatic, the splanchnics, the median, the pneumo-gastric and sympathetic, and the infra-orbital nerves. These were excited continuously for ten consecutive hours...

Claude Bernard himself came to realise the suffering involved in such experiments, and the Royal Commission heard his view that 'there are not strict enough measures put upon vivisection; that people rush to do it, and get the roughest results'.

I have dwelt on an example from Victorian England. Yet the basic criticism of vivisection does not rely only on examples of animal suffering. Firstly, it questions the reliability of animals as 'human models', suggesting that results obtained with animals may not apply to humans. Secondly, it condemns as wildly exaggerated the assumption that medical advance has been due to animal experimentation, claiming for example that improvements in sanitation and housing have been just as important. Thirdly, it doubts that the scientific community is honest when it claims to reduce animal suffering. As has already been pointed out, the new developments in transgenic animals suggest that more animals may be used in the future, not less.

4.4 Animal minds, human morals and the study of ethics

I hope that I have described fairly some of the arguments put forward by those who support, and those who oppose, animal experimentation. I hope too that the arguments show why this issue leads to such a polarisation of opinion, with two sides talking past each other, and condemning each other as mad or evil. Generally speaking, the so-called debate over vivisection gets no further than a repetitive declamation by each side, and little progress is made. Yet the subject is considered important by many people, both inside and outside the scientific community. There may be common ground that can be worked on, without shouting, but for that to happen we will need to stand back a little, and explore further how this moral problem can best be tackled.

Ethical problems are not simply those that concern actions by people that can be considered 'right' or 'wrong', 'good' or 'bad'. If you are thirsty you will want a drink, and will act accordingly. However this is unlikely to pose a moral problem, unless, for example in taking a drink, you deny someone else the possibility of quenching their thirst. The question of how we treat animals in science is one that requires a moral decision because we accept nowadays that animals can suffer during experiments. In addition, the decision is moral because, having committed ourselves to animal experimentation, we would tend to criticise as morally wrong, as well as criminal, those who would try to interfere with our actions in a laboratory.[5]

Our comparison of the arguments for and against vivisection suggests that both sides consider they are morally right, and their opponents morally wrong. One side argues that animal experimentation is morally right because (amongst other reasons) it saves lives; the other argues that animal experimentation is morally wrong because (amongst other reasons) it causes animal suffering. Left like that, the argument will exhaust

[5] Moral decisions are involved in the use of animals for providing food and clothing, in entertainments ranging from bullfights to circuses, and of course in scientific research. Those decisions will vary from person to person: you might eat meat, wear leather and bet on horses, but still consider that the testing of medicines on animals is wrong; or you might enjoy bacon and eggs, but still consider that there are right and wrong ways to rear pigs. It is important to realise that ethical decisions are not like mathematical calculations, for example many vegetarians wear leather shoes and swallow aspirin.

itself with accusation and counter-accusation, with no common ground explored. It is the job of ethics to take a sterile argument like that and revive it, take it forward so that at least we can see more clearly how the argument came to exist in the first place. I will try to keep our ethical analysis of animal experimentation reasonably brief. However, it must be admitted that there is a purpose to the apparently interminable deliberations of ethics. It is to allow an argument to be kept going for long enough to give it the chance of transformation from mud-slinging to genuine dialogue. In the case of vivisection, that dialogue is important.

One basic problem for ethics is in deciding – or discovering – the rules by which moral decisions can be made. For some it is a religious text that will provide the rules. For others it is the head teacher, or the views of friends or colleagues. It seems hard to find absolute, secure rules. For example consider the rule that prohibits murder; this is suspended readily enough in war. Yet though it may be hard to find absolute rules, it is not hard to find common moral beliefs. In spite of war and armies, killing people is generally considered to be wrong. What seems clear is that ethics must be understood as having a history. The different moral stances that people take on animals, therefore, can be understood not simply as a matter of their own personal beliefs, but as being part of a tradition. We will explore a number of such traditions, showing how they have been used to give moral guidance on how humans should treat animals. They are:

1   Animals cannot reason.
2   Animals have no awareness; they are simply machines.
3   Animals can suffer.
4   Animals have rights.

The first of these beliefs is the most ancient, dating back to the Greeks. The second belief is associated with the seventeenth century French philosopher René Descartes. The third belief became important in the nineteenth century, while the fourth is much more modern, dating from the 1980s.

Animals cannot reason
The issue of how we should treat animals is one of many philosophical questions that we can trace back to ancient Greece. Moreover, we see in Greek philosophy questions of whether animals could think, or could have a soul. It was the disputed existence of these qualities that exercised Greek philosophers. The mathematician Pythagoras once stopped someone beating a dog. He thought that he detected in the dog's yelps the voice

of a friend who had recently died; this suggested to Pythagoras that the soul of his friend was now residing in the dog. However, by far the most influential Greek voice was that of Aristotle, who lived slightly later, and was to have a huge influence on Western thought. Aristotle denied that animals can reason. This means that they cannot be said to have thoughts and an intellect, or beliefs and emotions. How would Aristotle have described the alert movements of a gazelle who believes that a lion is near, or the behaviour of a dog who greets his owner? While agreeing that these acts look like beliefs or emotions, Aristotle believed the signs to be purely superficial; there were no accompanying intellectual activities, nothing that could be put down to reason. Aristotle was impressed by the perceptual abilities of animals, but argued that their ability to sense and recognise things, to remember them and act appropriately, depends not on an ability to reason but simply on the proper functioning of the senses and the organs. Not everyone agreed, then or now, but as Aristotle was a major philosophical influence for 2000 years, many things followed from his position.

By denying that animals have beliefs, emotions and an intellect, Aristotle drove a wedge between humans and other animals (he knew that humans were animals); animals can do many things superbly well, but one thing they cannot do is reason. For Aristotle, this was an important deficit, because it justified something else he believed strongly. This was that animals are on earth for the use of humans. To put it more strongly, and using the teleological form that Aristotle favoured, the very existence of animals is to be explained by their usefulness to humans. Tame animals (oxen, sheep) are for food and for clothes; wild animals are also here for us – to be hunted and used as we wish. This teleological theme carried on in Greek philosophy, and led to some interesting ideas. One philosopher, Chrysippus, claimed that bedbugs exist because they are useful for waking us up. Mice make us tidy away old food: their existence is explained by their good effect on household orderliness. What was it though that made humans so special that the whole of life was organised around them? For Aristotle, our unique ability was reason, a capability more important than any other animal skill. Reason brings us close to God, and makes us superior to all other living things. This view was used to justify slavery too: slaves 'cannot reason' so they are naturally inferior, and are best governed by free men. As so often with ideas from ancient Greece, we find these themes running right through the history of ideas. When the theologian St Augustine discussed the commandment 'Thou Shalt Not Kill', with its potentially radical implications for how humans should treat animals, it was the Aristotelian denial of animal rationality that he turned to. St Augustine, of course, could also cite the Book of Genesis, which describes how, after the Fall of Man, God permits humans to eat meat. There are other stories too in the Bible, which suggest that animals are 'for' human use: Jesus blasts a tree because it did not bear fruit, and he caused a herd of pigs to drown itself in the sea. However, the Bible also contains plenty of references to harmony amongst animals, or between animals and humans, and so for a philosophical position, St Augustine turned to Aristotle, and to the denial of reason.

Though dismissing the idea that animals can reason, Aristotle did not thereby belittle their skills. In fact, he wrote a good deal about how it is that animals can be so skilful (for example in hunting) without actually doing any thinking. It would be a scholarly task, far beyond the scope of this book, to analyse Aristotle's views on animal behaviour,[6] but we see today an analogous issue: is it permissible to talk about the behaviour of animals as though they are, in some sense, like humans? When we talk about the behaviour of animals, we often use words like wait, search, and choose. For example if we wanted to describe lions hunting wildebeest, we might well say that the lions searched for a herd of wildebeest, waited for the right moment to attack, assessed the situation, and chose a weak animal to hunt down. Such language is described as 'anthropomorphic', the term used when a non-human phenomenon is described in human terms. It is customary for biologists to question the legitimacy of anthropomorphic language on the simple grounds that a lion is not a human. However, it is equally customary for biologists to use anthropomorphic language; in fact, it is essential. Of course, some anthropomorphic language is indeed ridiculous. If I say that my cat is off his food because of last night's football result, then I am joking – or daft. My cat does not have views about football, and to attribute his food problem to a football result is plain silly. However, just because some anthropomorphism is daft does not mean that all anthropomorphism is daft. In the case of the lion hunt, those terms suggesting that the lions have a game plan serve our purposes very well. There does not seem anything much wrong with saying 'The lion waited in the long grass.'

[6] *Animal Minds and Human Morals* by Richard Sorabji (Ithaca, N.Y.: Cornell University Press, 1993) gives an in-depth but highly readable and interesting analysis. For a sense of how philosophers discuss ethics refer to the anthology of readings *A Companion to Ethics* by Peter Singer (Oxford: Blackwell, 1991), *Ethics and the Limits of Philosophy* by Bernard Williams (London: Fontana, 1985), and *Ethics: Inventing Right and Wrong* by J.L. Mackie (Harmondsworth: Penguin, 1997).

Is it unacceptably anthropomorphic to suggest that lions engaged in hunting, or waiting, have beliefs? According to Aristotle's philosophy, yes. For to say that an animal is waiting is to suggest that it has a belief about the future, which is reasoning. To explain how animals can have such complex behaviour, and yet have no beliefs or power of reasoning, Aristotle argued that they must have very advanced perceptual systems. He systematically dissected animals, and observed their behaviour, with a view to explaining how an animal can sense something, seem to remember it and act on it, without actually having to reason. Perhaps Aristotle is right, and lions cannot think or have beliefs. Then our anthropomorphic language is simply a short cut, and nothing else. Perhaps it would be better if 'the lion waits in the grass' was translated into a sentence, where there are no words like 'wait'; but to do that would be to produce long-winded and obscure terminology about, amongst other things, motor neurones and sensory systems and muscles, all of which seems a long way from capturing the drama – and the reality – of the lion's behaviour. More importantly it would be to reject entirely the idea that a lion can reason and can 'wait' in the same way that a human can 'wait'. So far, science has not shown that a lion cannot reason. Indeed, there are some grounds for being cautious about rejecting the idea that lions can reason. Lions share with humans a long evolutionary trajectory, and a similar nervous system.

The problems over what is meant by such terms as reason, mind and consciousness have traditionally been the work of philosophers, not biologists. Biologists may use anthropomorphic language in discussing animals, but in the main have ignored asking questions about animal minds. Darwinism has made this possible by seeing many interesting biological questions in the adaptedness of the lion's hunting strategy (or any behaviour of any animal) rather than in its thoughtfulness. Until recently, animal consciousness was barely considered by biologists, however anthropomorphic their language. That is now changing, as developments in neurophysiology and in computing finally bring the study of consciousness into the ambitions of biologists. Moreover, consciousness in humans and in other animals must have evolved, suggesting that it is adaptive and has survival value. It would be counter-intuitive to argue that consciousness has survival value for humans, but not for other animals. Arguments like this may well come to show that our anthropomorphic language is not a short cut at all, but is a legitimate reflection of mental similarities between us and other animals. If that happens,

Aristotle's influence will wain, and our views about our treatment of animals will have to be revised once more.

Animals are simply machines

Aristotle did not discuss cruelty to animals. While he was clear that animals are for humans, and we can do anything we like with them, he did not specifically address what might have seemed obvious, namely, the question of cruelty. Perhaps he thought that an animal that cannot reason cannot be a victim – any more than a rock can. Certainly the idea that animals have no mental life was to be used to justify a great deal of cruelty. To frame this discussion, we will look at two Frenchmen, both figures from history, but who lived more recently than Aristotle. They are the essayist Montaigne (1533–92) and the philosopher Descartes, who lived slightly later, dying in 1650. Both were well aware of ancient philosophy. However, they took the argument further. They were interested in how the alleged lack of reason in animals might have implications for human behaviour. Montaigne did not consider the implications for science, but Descartes is a major figure in the arguments over vivisection.

Michel Eyquem de Montaigne was a French aristocrat and essayist. He wrote about a great many things, and liked to pride himself on his freedom of thought, and on his eclecticism. Amongst his essays you can find titles as diverse as 'Of Drunkenness', 'Of Thumbs' and 'Of the Education of Children'. Montaigne wrote during the Renaissance, a time in Europe when scholars were beginning to challenge the old traditions handed down by Church thinkers, who as we have seen, were much influenced by Aristotle. The Renaissance is the era of Michelangelo, Leonardo da Vinci, Galileo and Machiavelli; it is the time when printing became common, and when towns began to set up universities. Travel – of goods, people and ideas – became easier. Most important of all, science got under way, and started analysing nature experimentally, rather than relying on the textbooks.

Montaigne liked to write about himself and his feelings. He said 'I have never seen a greater monster or miracle than myself.' He felt it quite reasonable to write about his opinions, and was very sceptical of established authority. Amongst the traditions he saw as plainly false was the notion that animals have no reason, and that we are superior to them. He argued that we have obligations at least to some animals, and wrote in his essay 'On Cruelty': 'I have never been able to see without discomfort even the pursuit and killing of innocent beast, who is without defence and from whom we receive no offence.' He went to some lengths to criticise the idea that animals cannot reason. He argued that there is communication between different species of animal: 'by a certain barking of the dog, the horse knows that he is angry; by a certain other tone of his, he is not startled. Even with the beasts that have no voices, from the interchange of services that we observe between them we readily infer some other means of communication; their motions converse and consult'. Moreover, he saw in the animal life around him all the signs of reason and awareness. When Montaigne notices a swallow looking for a good nesting place in the roof, he sees an animal making a choice and showing judgement. He objects to the Aristotelian idea that such skills are independent of reason: 'there is no ground for thinking that the beasts do by innate and enforced inclination the same things that we do by our choice and skill'. Montaigne comes to the conclusion that it is folly to imagine that people are superior to animals. There may be differences in intellectual power or in imagination, but they do not justify the view that animals have no reason, or are entirely distinct from people.

All this was quite subversive, but the ideas that Montaigne attacked were reasserted when Descartes concluded that animals cannot reason. Descartes did not trust his teachers. He describes how, during the time he was a soldier, he took refuge from the battle, and tried to forget everything that he had been taught. Hidden and free from distraction, he asked himself, 'Can I be sure of anything?' He asked himself whether other people exist, and admitted that he could not be sure: we encounter other people vividly enough in dreams. Alone in his refuge, Descartes found that he could not be sure of very much at all, except for the fact that he, Descartes, existed. He knew that because of his own consciousness. By being able to question 'Do I exist?' Descartes demonstrated to his own satisfaction that he, at least, existed: 'I think, therefore I am' he announced, presumably to himself. From that position of security, Descartes then builds up a system of inquiry that in his view will provide a strong basis for science and philosophy. Thus Descartes used a method of doubt to establish a foundation for truth; he did not rely on religious or Aristotelian authority; he showed how thinking can provide a sure way of getting through life. Descartes is regarded as the founder of modern philosophy, and an important influence in the development of science.

What does all this have to do with animals? Descartes was a religious man, believed in God, and knew from the science at the time that animals and people are anatomically similar, but like many before and since, he wanted a way of splitting people from other animals. He argued that animals are different because they have no immortal soul; for our purposes, however, the significant denial he made was that animals are not aware of anything at all – they are automata, nothing more than very complicated machines. On this view, animals can neither feel pain, nor suffer generally. Descartes was aware that this view would be seen as extraordinary and against all the facts. However, he said that peoples' tendency to see animals as having thoughts and emotions was a prejudice left over from childhood. Animals that seem distressed, or that leap away from painful stimulus, or rush for their food, are simply behaving automatically. They are alive, they have complicated mechanisms for sensing the world, and for acting, but they cannot actually feel anything in a way analogous to the human experience: in short, they have no consciousness. Animals sense an injury, and struggle and cry, but these are no more than the creaks and rattles of a broken machine. This quote, from one of his contemporaries, makes it clear what Descartes meant, and what implications his beliefs might have for animal welfare: 'The scientists administered beatings to dogs with perfect indifference and made fun of those who pitied the creatures as if they felt pain. They said the animals were clocks; that the cries they emitted when struck were only the noise of a little spring that had been touched, but that the whole body was without feeling.'[7]

Remember that Descartes's purpose in mind was to separate man from animal. He had said, in another context, that a human body is really only a machine, with nerves and bones replacing cogs and gears. The part of the human that can think, according to Descartes, is separate from the body, is called the mind, and is immortal. This demarcation has its own philosophical problems: if the mind really is separate from the body, how does it connect with the body, as of course it must, for example when we blush, or make a voluntary movement, or speak? The splitting of mind from body, the idea credited to Descartes, is called the 'mind–body problem', or 'dualism' and has generated a huge amount of philosophical debate (it is also discussed in Section 2.7, in the context of the philosophy of reductionism). Descartes was enthusiastic about dualism – the separate mind – because here was a device for separating animals from man, as least so long as you can show that animals do not have a mind, whereas humans do. An important aspect of his evidence was that while humans have

---

[7] The quote comes from Tom Regan's *The Case for Animal Rights* (Berkeley: University of California Press, 1983), p. 5.

language, animals do not (he discounted as absurd Montaigne's examples of parrots and jays that seem to talk). Aware that his views on animals might be objected to, he said 'My opinion is not so much cruel to animals, as it is favourable to man.' Yet though we might find that Descartes is wrong to deny any possibility of consciousness to any non-human animal, nonetheless we see some interesting historical continuities between the way he talked about animals, and the way that we do. When Descartes wrote 'doubtless when the swallows come in spring, they are operating like clocks', he sounds simplistic. Yet it is not hard to find the terms 'biological clock' or 'mechanism' in a modern textbook.

Through our brief discussion of Descartes we begin to see the basic shape of the dispute about how we should treat animals. In terms of the capacities of animals, the concepts likely to be considered relevant are the awareness of pain, the possession of consciousness, and the possession of language. We can see that modern knowledge about animals might itself inform the debate. Can we use neurophysiology to discover whether an animal thinks like us? When an animal reacts to what we would ourselves consider a painful experience, can we really not know whether it is feeling pain? When we say that no animal has language, how is it that researchers have been successful in communicating with chimps? The questions multiply: what do we mean by thinking, reasoning and language, what is consciousness and, for that matter, what is pain? These words are particularly hard to define, and this is one reason why the scientific study of consciousness is so difficult. How can you measure or test something that you cannot even define? In all of these questions, there is something implied: our treatment of animals should be influenced by our understanding of animal consciousness.

It is not easy for one human to find out what is going on in the head of another human, even if we talk to each other. If you say you feel sad, I can empathise because I know what I feel like when I feel sad, but I cannot experience exactly what you are experiencing, nor even know for sure that we see colours or hear sounds in precisely the same way. How much harder it is to find out what is happening inside the mind of a lion or chimp – if we are correct to talk of such animals having minds. The Cambridge philosopher Wittgenstein (1889–1951) said 'If a lion could talk, we wouldn't understand him.' Others have replied 'If a lion could talk, he wouldn't be a lion.' Others still have questioned both these ideas. Why could we not have a reasonable communication with a lion, perhaps discussing whether we are hungry or not, or what the weather is like? The point is that the problem of determining the mental life of an animal is difficult because of both definitions and the techniques of study used.

Animals can suffer

Most scientists, like most people, believe that animals experience pain, and can suffer. Whatever the philosophical problems involved in being clear about what someone else, or something else is experiencing, common sense might suggest that it would be perverse to deny pain to animals. After all, the sensory and motor systems, the synapses and neurotransmitters, are basic to the vertebrate world. Brain size differs, of course, but that does not in itself rule out the notion that animals have pain. Most important of all, animals react as though they are in pain. Science then, as well as common sense, seems to disagree with Descartes (or at least with the standard interpretation of his writings). Similarly, it seems perverse to deny that animals can suffer. To suffer is to have your well-being seriously harmed, for a length of time. It is to have a preference, or a need, denied. Pain can cause suffering, but so can other sorts of harm, such as overcrowding, or lack of food, or lack of a mate. Suffering, then, seems a broader kind of experience than pain. If we agree that animals can feel pain, we will take the signs of an animal in pain seriously. Those signs, while different for different animals, can probably be determined, and even used by legislators, but can suffering be measured?

Clearly, definitions will once more be important. The turkey farmer at Thanksgiving or Christmas, talking about his intensive farming, is quick to point to the warmth, the antibiotics and the food he guarantees every bird. Indeed it is not in his interest to have the turkeys dropping dead from disease, or starvation; instead, they get fat quickly. The crowded conditions, he says, do not in themselves cause suffering: turkeys simply do not mind, they are too stupid. A very few turkeys may be unlucky, and get trampled on, or miss out at the grain bin, and these will suffer; but the majority does not suffer, and so the turkey farm cannot itself be said to cause suffering. Yet work on chickens shows that if we allow that chickens have preferences, then more subtle measures of suffering can be used, and can give an insight into the experience of an intensively farmed animal.

Chickens, if given the choice, prefer a dusty, messy floor to the kind of grill floor found in farms. That is, if you give a battery chicken a choice between two boxes of the same size, where one has a grill floor, and the other a solid floor with grit and dust, they choose the latter; the type of cage matters. Therefore, chickens, even the lines bred generation after generation for intensive farms, still have some of the inclinations of their wild ancestors, who spent a lot of time hopping around on forest floors. These inclinations were investigated by Marian Stamp Dawkins in a series of classic experiments during the 1980s. The question was asked 'What are the desires of chickens?' A seemingly strange question was transformed into a meaningful scientific exercise by giving egg-laying chickens a choice of cages. When each cage is similar in that food is visible, but differs in terms of what the floor is like, and how big the entrance is, chickens will opt to squeeze through a narrow gap to get into a cage where the floor is gritty. In such circumstances, they ignore the cage that has food, and an invitingly wide entrance, but also has a grill floor. Chickens do not like squeezing through gaps, so the fact that they will undergo that inconvenience in order to be in a gritty floor cage tells us something of the 'sacrifice' that they are prepared to make. Clearly, by varying the gap, and the floors, a great deal can be learnt about how high a sacrifice a chicken is prepared to make in order to get the cage she likes. Similar experiments can be done with other farm animals. Pigs, for example, learn to bash a metal plate if they are rewarded with the chance to have a snout-to-snout nuzzle with a neighbour, but if the reward is food, it is found that pigs will plate-bash almost ad infinitum in order to get the food. Pigs like nuzzling, but in the context of this experiment, they like food more.

These animals, then, have preferences – or is such a statement an anthropomorphism? One could be even bolder, and say that these animals have emotions. The pig that persistently bashes a plate in order to get food, cares about that food and wants it. The chicken that ignores the cage with a 'barn door' of an entrance, but squeezes through a 9-cm aperture so as to get in the cage with a gritty floor, cares about its environment: are these not emotions? An alternative way of putting this would be to say that the choices the animals make are not choices at all, but are simply the triggering of innate behaviours, or the learning of conditioned behaviours. This philosophy of reducing complex behaviour down to a simple terminology, where everything is a 'response' or, worse, an 'output', is known as behaviourism; it is much less common now than 50 years ago. Part of the problem of behaviourism was that it seemed to imply that human behaviour itself was simply a set of 'responses'. When the question is asked 'Can an animal think?' the behaviourist might answer 'No, and nor can a human.' Now that behaviourism is less fashionable, it is much more acceptable to hypothesise that pigs, as well as humans, have emotions.[8]

What about language? The idea that animals cannot talk is important in the history of the relation between humans and animals. Each of the authorities we have mentioned (for example Descartes and Montaigne) have argued about this issue. Once more, problems of definition mean that we can only discuss the problem superficially. Definitions are important because we need to distinguish whether we mean communication, or talking, or signalling. As a rough guide, we could say that cells signal, ants communicate, but humans talk. Books could be written, and are, assessing the reliability of these distinctions. Similarly, many scientist-years have been spent investigating whether animals – for example parrots, or chimps – can have language. The drive behind this interesting work has partly been the twentieth-century intellectual orthodoxy that a thought or a concept can only exist if there also exists a word to describe it. In other words, language precedes thoughts and beliefs. This is almost the same as saying that you cannot have consciousness if you cannot talk; if true, then animals do not have awareness, feelings, emotions and beliefs, however defined.

Chimps cannot learn to talk, but they can learn sign language. Chimps, to a certain extent, can string the signs together to make sentences. Yet the sentences are simple, very concrete ('hungry apple now'), and rarely spontaneous. When chimps sign, they do not get very far. Once again, when teaching chimps to signal at least as much effort has to be spent in getting your definitions right as on the actual teaching. It is very easy to think that an animal is being cleverer than it really is, as shown by the following story about Clever Hans. The point about Clever Hans is that though he was a nineteenth-century horse living and working in Germany, the problem he illustrates is very hard to overcome, even today. Hans gave public demonstrations at which he showed how well he could add. A member of the audience would ask a question ('What is four plus three?' for example) and Hans would give the – invariably correct – answer by tapping his hoof the appropriate number of times. Hans was so successful, and his owner was

[8] These experiments are described in *Through Our Eyes Only? The Search for Animal Consciousness* by Marian Stamp Dawkins (New York: W.H. Freeman, 1993). For a consideration of how the existence of animal consciousness implies the existence of animal rights, see Tom Regan's radical text *The Case for Animal Rights* (Berkeley: University of California Press, 1983). Stephen R.L. Clark's *Animals and Their Moral Standing* (London: Routledge, 1997) is a collection of papers covering many issues concerning animal welfare and morality.

making so much money, that German scientists ran their own investigation. A student, Oskar Pfungst, looked into the matter, and realised that, somehow, the horse was responding not to the question, but to his owner. This was proved by asking Hans to do a sum, and giving the owner (who was convinced of the genuineness of Hans's ability) another one. Hans then got the answer wrong. He could only get the answer right if his owner had it right too. Now it was clear what was going on. When a member of the audience shouted out a sum, Hans started tapping the floor. As soon as Hans reached the correct number of hoof taps, his owner made, quite involuntarily, a tiny movement of the head, and the horse stopped tapping. The interesting point about the story is that the owner was quite unaware of this twitch on his part. Moreover, the sensitiveness of Hans (and all his experimental descendants) meant that even Oskar Pfungst, who know exactly what was going on, could not stop himself from giving involuntary twitches either when Hans was asked to count. If Pfungst knew the correct answer, and Hans could see Pfungst, then Hans tapped out the right number.

Though it is interesting to study language in animals, such studies will not necessarily give us guidance on how we should treat them. Whether an animal has syntax should have no moral consequences whatsoever. We can say this with some confidence, at least if we believe that animals (or some of them) can have emotions, beliefs and desires, even if they do not have language. However, there is also another reason, perhaps more basic, and it is best put by the eighteenth-century philosopher Jeremy Bentham who in 1823 asked, famously: 'The question is not, Can they reason? nor Can they talk? but Can they suffer?' With that question posed, we can move on to two anti-vivisection arguments existing today. In the chapter so far we have found that of all the ideas that might influence us in our behaviour to animals, two seem most relevant. The first idea, first properly articulated by Bentham, but accepted by the scientific community, is that animals can suffer. The second, more controversial because more difficult to investigate and define, is that animals have preferences and emotions.

Animals have rights

One important argument against vivisection starts with the observation that animal experimentation causes suffering. Here the scientists and the anti-vivisectionists are likely to be in agreement, even if they argue over the extent of the suffering. The divergence comes because, for the scientific community, animal suffering is less important than human suffering, and so it is legitimate to allow animals to suffer, if human suffering is lessened.

In his book, *Animal Liberation*,[4] Peter Singer describes such views as 'speciesist', and condemns them as being just as evil as sexism or racism. These familiar social pathologies are acts of discrimination against people simply on the basis of their gender or their colour or ethnic origin. An enlightened society considers such discrimination wrong, and believes so on the grounds of an even more fundamental belief, namely, that all humans are equal. This is not to say that all humans are the same, or should at all times be treated the same. Rather, certain important basics, like access to education, or hospital treatment, or pay, should not be determined by gender or race. According to Singer, speciesism is the practice of discriminating against a being simply on the grounds that it is not of your species. In the case of vivisection, the speciesism consists of believing that the suffering of a rat, or a rabbit, or a monkey, is less important than the suffering of a human. This speciesism, according to Singer, is immoral in just the same way that sexism or racism is immoral. The suffering of a rat has the same worth as the suffering of a human; or to put it another way, it is immoral to believe that the suffering of a rat is acceptable simply because it is a rat.

If we were to consider animal suffering as important and as acutely felt as the suffering of a human, then the way we treat animals would change radically. At least, that is Singer's view. *Animal Liberation* contains many examples of animal suffering that most scientists would, I suspect, condemn. As a result of Singer's emphasis on taking animal suffering seriously, there is some possibility of dialogue between his views, and those of the vivisectionist. For, as any scientist will tell you, science – and the laws that control it – also takes the suffering of experimental animals seriously, and aims to minimise it. The US Animal Welfare Act takes if for granted that experimental animals can suffer. In that Act we find a requirement that 'animal pain and distress are minimised [by the use of] adequate veterinary care with the use of anaesthetic, analgesic, tranquillising drugs or euthanasia'. The Act requires that the 'principal investigator considers alternatives to any procedure likely to produce pain to, or distress in an experimental animal'. When a procedure is likely to cause pain then the Act requires that a vet be consulted, and that the researcher makes provision for 'the pre-surgical and post-surgical care by laboratory workers'.

ANIMAL MINDS, HUMAN MORALS AND THE STUDY OF ETHICS 135

Nevertheless, the Act does not want to prohibit painful experiments entirely. Consider this: 'the withholding of tranquillisers, anaesthesia, analgesia, or euthanasia when scientifically necessary shall continue for only the necessary period of time; that no animal is used in more than one major operative experiment from which it is allowed to recover except in cases of (i) scientific necessity; or (ii) other special circumstances as determined by the Secretary'.

Clearly the Act is trying to control the suffering of animals, by calling for the use of painkillers and appropriate 'post-surgical care' and/or anaesthesia. However, the cool tone of the Act shows the respect given not only to the animals, but also to 'scientific necessity'. Scientists in both the USA and the UK have to acquire a license when they wish to perform animal experiments, and the acquiring of this license demands that a great deal of information is put forward to the authorities. This, indeed, is what researchers point out: that they are required by the law to adhere to standards, standards that they in all likelihood approve of. How would an anti-vivisectionist view this Act? They would note with some queasiness the invoking of scientific necessity. No doubt it is the scientists who are the judge of that. Still, there are regulations, and animal suffering has been recognised as real. Researchers are required to do their utmost to reduce the suffering by anaesthesia, to replace animals with non-animal models, and to use as few animals as possible.

Singer has been criticised for basing his argument on the utilitarian method of deciding moral issues. According to utilitarianism, what is important when such decisions are made is the net effect of the decision on the happiness, or pleasure, of everyone effected. A decision is a morally good one if, overall, it enhances pleasure, and reduces pain. The attractions of utilitarianism are that it seems to offer a simple way of making a decision, based simply on the consequences of an action. There is no need to refer to God, or laws, or tradition. However, in the attractions of utilitarianism lie its downfall. For how do you measure the pleasures caused by your decision, or the pain? When NATO bombed the Serbian capital Belgrade in 1999, the UK Foreign Minister was interviewed one morning on the BBC, just after a bombing mistake had killed three people in the Chinese embassy. Questioned about the mistake, the latest of a dozen, the Foreign Minister justified the overall campaign, and its inevitable targeting slip-ups, on the grounds that, overall, it would help to restore peace, and enable the Kosovan people to live peacefully in their villages. The BBC interviewer retorted that the argument does not work if it is your mother who has been killed in one of those bombing mistakes. That is the problem with utilitarianism: it aims to make equal, and then to quantify, the pleasures and pains of all people, but when it comes to making decisions that affect others, most people want to know who those others happen to be.

Peter Singer's argument about speciesism is deployed so that animal and human pains and pleasures are given equal weight. Even if this is accepted, the problems of utilitarianism assert themselves. Consider the case of a drug company developing a new drug for soothing headaches. To decide whether such testing is morally acceptable, using the utilitarian viewpoint, one would have to compare the suffering and death of the experimental rats against the mild discomfort of people with the headaches they have developed on the drive home. In some cases, no doubt, the calculation results are obvious enough: for example cosmetic testing on animals has been banned in the UK. The balance may be more difficult to weigh, or may even go against the animals, when we consider the case of properly treated animals, anaesthetised and never allowed to suffer, used to develop new coronary surgery. Peter Singer's view depends on balancing pleasure and pain. The criticism goes: is the calculation possible?

Nor surprisingly, there is a more radical view. According to the animal rights argument, an animal has inherent value by virtue of the fact that it has beliefs and inclinations. In this view, an animal has some idea of the future, it has beliefs, in short is has a psychological identity. The details of that consciousness, and its similarity to that of humans, may be disputed, but cannot be altogether rejected. Indeed, according to the animal rights view, it makes sense to credit animals with a psychological identity – a sense of self – and it is this that gives animals certain rights, in exactly the same way that we accord certain rights to humans. Just as humans have the right not to be killed, imprisoned or tortured, so do animals. For the point about rights is that everyone has them, simply as a consequence of being born a human being. If animals have an inherent value and a psychological identity, then the same rights should be applied to them.

This is more radical than the previous argument. For now there is not much relevance in the fact that an animal is well watered and fed, anaesthetised and quickly and painlessly killed at the end of an experiment. A utilitarian – someone who measures consequences – might be reassured by seeing a farmyard of free-range pigs, or the comfortable cages of oncomice. A proponent of animal rights would not, for the pig will be slaughtered, even though its own desire would be to go on living, and the oncomouse will be the subject of experiments that will cause it harm (for untimely death is a harm) even if it never suffers any pain. The essential difference between the utilitarian view and the rights view can now be seen. The utilitarian view wishes to protect animals by giving the sufferings of animals as much importance as the sufferings of humans. It makes moral decisions by considering the consequences of an act. However, from the animal rights view, it is not consequences that are the prime consideration, it is the inherent value of the individual that is important. This value is acquired by virtue of an animal's consciousness: its ability to have thoughts, beliefs, desires, memories, and a sense of the future – as well as feelings of pleasure and pain.

4.5 In conclusion

Moral philosophy sometimes suffers from sounding unreal: texts on the subject often discuss cannibalism and murder as though these are all part and parcel of an undergraduate career, and their proper comprehension intrinsic to university life. A biologist, in reading the ethics discussion above, might find some of the arguments unconvincing for being unreal, for being unscientific. Perhaps the first unease comes from the universal use of the word 'animal', when what is meant is 'mammal'. In this chapter I have also used the word animal in that sense. However, there are two, more important, problems. One is to do with the facts of animal consciousness; the other is to do with deciding the cut-off point where we can agree that an animal is so 'lowly' as to have no thoughts and beliefs and, therefore, according to the two anti-vivisection philosophies argued under 'Animals have rights' (Section 4.4), can neither suffer nor have rights. Let us take a look at this problem, the so-called 'slide argument'.

The fact is, consciousness is extremely hard to study. I have described the problem earlier as a problem of definitions. If you want to find out whether an animal has emotions, or thoughts, or desires, or beliefs, you first have to be clear what you mean by those words. Even if you can define these words, then the second step, testing the presence of what they stand for, may well be difficult, especially if the subject has no language. Thus, while consciousness is something we all know a lot about in our lives, and is something that can be explored properly by poets and artists, scientists have been much more cautious. So if consciousness is to be at the root of our decisions on how to treat animals, then our problems in studying it will make those decisions equally difficult. For example, our treatment of frogs might come to depend on what we know about the minds of frogs.

It may seem absurd to talk of the mind of a frog, but it does not seem absurd to talk of the mind of a chimpanzee. Where on the spectrum of animal life does consciousness begin? According to the slide argument, the fact that there can be no cut-off point means that it is unfair to accord rights to some organisms, but not others. The slide argument, therefore, is a *reductio ad absurdum*: it argues against giving rights to chimps, or rats, because by deciding against the case of the earthworm, or a locust, we are being arbitrary. Yet it is clearly absurd to give an earthworm the same rights as a human, or to claim that the sufferings of a worm are as important as the sufferings of a rat. In response to this, commonsense can be applied. While it is true that we do not know much about the mental activities of a frog, nor even how to define them, that does not mean that we should dismiss as incoherent the idea that mammals have a mental life. Perhaps as science and philosophy works out better how to investigate the understanding of animals, reliable criteria for measuring it will be found. Until then, we should not suspend the legal framework that already exists for protecting animals, nor simply rule out the possibility of consciousness in all non-human animals.

However, we must beware of thinking that the new understanding of consciousness will resolve all the ethical issues currently surrounding our use of animals. Facts on their own cannot decide a moral problem. Thus, we might one day construct a reliable idea of what it is like to be a rat; we might even be able to empathise with a rat. Yet even if we can get inside the mind of a mouse, we might still choose to run experiments on it because we consider the pursuit of knowledge worth the sacrifice of the mouse. In philosophy, the error of thinking that facts will resolve a moral problem for you is called the naturalistic fallacy: put simply, you cannot derive an 'ought' from an 'is'. Thus, you cannot decide how to treat animals simply on the grounds that they are conscious. You need to make a value decision as well, perhaps that conscious creatures have the same fundamental rights as humans, or that the needs of humans outweigh the needs of other mammals, however profound their consciousness.

Biology will always have ethical problems. We have looked in detail at just one, and tried to unpick the various factors at work. As regards vivisection, your own views are for you to decide. My purpose in the chapter has been to suggest that biology involves moral decisions, which in turn demand an appraisal of your own values. Biological research will not on its own make the decision for you. This is why some knowledge of ethics is vital for a biologist. These are decisions you will have to make for yourself, and they will go beyond the facts available to you. You will have to depend on your own ability to reflect, and to act. There is every sign that the biology of the next few decades will be marked by a huge increase in the importance of ethical debate. Such debates will require personal involvement, and will be exciting. They will not go away simply as a result of scientific developments; on the contrary, they will multiply.

Controversies in biology

5.1 Controversies in biology – and in the media

History, and the changing face of scientific research, tell us that final truths are not the product of science. A consequence is the tendency for scientists to talk in terms of uncertainties and probabilities. There are those in research who talk up their findings, and amplify their significance. This is the daily rat race of competitive science. Yet it would be an unwise scientist who staked his life on every aspect of a favoured theory. In 50 years' time, or even in 10 years, laboratories will be different, and moving on. Of course, some ideas in science do seem rather stable. It would be a surprise to learn that the sun goes round the moon, or that DNA is a triple helix. In the active areas of science, however, where scientists pick over the minutiae, hypothesis and uncertainty are, quite simply, the rule. Nor does this seem to trouble scientists. Perhaps it makes their work seem more of a game, however competitive. Yet outside the laboratory, and especially when scientific topics are debated in the press, the evasive nature of science seems a puzzle. The popular image of science is not of a quizzical subject, but of an endeavour that will tell everyone what is going on, and what they should believe. This chapter looks at the way this misunderstanding is played out in three controversies. Each is newsworthy. Each displays in full measure the uncertainty of science. Each, too, shows how news, and science, are tangled up, and each seems to resist all attempts at 'closure'.

My first case study is a look at the way that scientists, politicians and the public reacted during the 1990s to the crisis over British beef, when it was found that mad cow disease (bovine spongiform encephalopathy or BSE) might be transmitted to humans simply by eating a meat pie. The disaster of BSE destroyed the livelihoods of farmers, cost millions of pounds, and made many politicians look very foolish. Throughout, there was a tension between the politicians' desire for certainties, and the scientists' desire to be faithful to the ambiguities and imponderables of the available evidence. My second case study is drawn from the debate over genetically modified (GM) foods. Again there are concerns over whether science can ever declare something 'safe'. As we shall see, the GM food issue has aspects that set it apart from the BSE crisis. For example pressure groups were highly significant in making the public aware of possible problems with this new technology. Moreover, whereas BSE was a crisis that could have been avoided, and was a British disease caused by a British decision, GM foods rely on techniques developed worldwide by scientists and promoted by multinational companies.

My final case study will perhaps be most fully debated in the years to come: xenotransplantation. The prospect of transplanting animal organs into humans appals some, but is eagerly championed by others. Part of the context of this controversy is the long wait usually endured by someone in dire need of a fresh heart, kidney or liver. Pigs' organs, genetically modified so as to reduce the chances of rejection, might be an obvious solution. Of the many objections to this strategy, one will form the focus of our debate: the safety of xenotransplantation. Once again there are uncertainties about this, which are quickly picked up by campaigners and politicians opposed to the practice. Once again biologists will face pressure to give a definitive answer to the politicians looking for certainties about the safety of xenotransplantation. We shall see that those who campaign against the practice point to other ways by which we could improve the cardiovascular health of the nation. I include the topic here because xenotransplantation also raises the kind of ethical issues that will soon emerge as machine intelligence gains in power. While bioethics is currently preoccupied by genetics, this will change in the first years of the millennium. The new debate will be the interface between people and machines, including our attitude to the implanting of intelligent, but non-organic devices, inside our heads, our guts or our limbs.

One unifying feature of my case studies is that they regularly enter the news. Another is that they involve the concept of risk. We shall see that it is hard to produce a public discussion of the risks of a scientific procedure, and that the problem concerns scientists and politicians, as much as newspapers. Newspapers and television channels are interested in food safety issues; they like stories about dangers lurking in things that we take for granted – lunch, for instance. The media has a thirst for stories that can be told quickly and simply, and for scoops and breaking news. Those commercial pressures bearing down on the media strongly influence the way that editors sort through the science ideas passing across their desks. It is an oversimplification to accuse the media of simply 'trivialising' science, and the public of misunderstanding it. It is more helpful to see that the priorities of newspaper people, and of scientists, are different. This mismatch of priorities is clearly seen in the way that science stories tend to be reported. Either there is an excited emphasis on 'breakthrough discoveries', in which case busy journalists are usually happy simply to take in one mouthful the press releases put out by publicity-hungry laboratories, universities and biotechnology companies; or the emphasis is on science 'out of control' or, at least, out of its depth. The idea that scientists are not fully in control, and do not quite know what it is they are unleashing on the world, makes for excellent copy, whether in tabloid headlines or earnest broadsheet leaders. The scientific community prefers the ordered stories provoked by press releases, and dislikes what it sees as the ignorance and sensationalism of the press coverage of, for example GM crops. The well-trained scientist needs to be aware of these two aspects of the public reception of science, and might look for a new synthesis in science communication, where both the stage-managed debate, and the violent bear fight, are avoided.

Because they are experts in a field of great public importance, biologists are asked for comment and advice. Often, such enquiries come from the press or from the politicians. In both cases, scientific reticence may not be what is required. More likely, both the politician and the news desk want a simple certainty, not an estimate. In the three case studies that follow, we see how the wavering uncertainties of the honest life scientist cause confusion in the world outside academia, and lead to spectacular misunderstandings. BSE forms my first case study, but the debates over the safety of genetically modified crops, and of xenotransplantation, further illustrate the point. In the case of BSE, the political desire for certainty was to prove very costly, both in terms of money and lives.

5.2 The crisis of 'mad cow disease'

In the 1990s it became clear that beef farmers had a problem. A new disease, invariably fatal, had arrived in their herds. The cattle affected became drowsy and found it hard to walk; eventually paralysis set in and they died from respiratory failure. Yet the disease could not be diagnosed, except by post-mortem. Animals that died from this disease had anomalous 'spongy' brain tissue; the name of the disease, bovine spongiform en-cephalopathy, simply states the fact that it is a disease of cattle in which the central nervous system develops holes. For a while the animals that died from this disease, or were slaughtered to prevent a lingering death, were sold to the plants that make animal feed. However, before long, a closer scrutiny began to be applied to the disease. Not only was it spreading remarkably rapidly (up to 4000 new cases a month by 1993), it was also clearly the bovine equivalent of the sheep disease scrapie, which has been common for hundreds of years. It was reasonable to suggest that the practice of feeding sheep tissue to cows had allowed the infectious agent to cross into cows. The question for biologists, epidemiologists and politicians was: would it jump twice, could it cross into humans?

In terms of biology, this is a tale of three diseases: BSE, Creutzfeldt–Jakob disease (CJD), and scrapie. The diseases have a number of things in common, beyond being fatal. They have a long incubation period, and have symptoms that are a consequence of the degeneration of the central nervous system. They also have a unique infectious agent, the prion, a particle devoid of nucleic acid and consisting simply of an abnormal protein. Prions are mysterious, but they became something more than a scientific oddity when it was realised that the prion that causes BSE could pass into the human food chain, and kill someone who had done nothing more dangerous than eat beef. When new variant CJD was judged to be a human form of BSE, contracted by contact with cattle, the beef industry was immediately faced with ruin. A European ban was imposed on imports from the UK, and the bankruptcies began.

Lizzie Fisher, a prion expert, gave a sense of the future when she wrote in 1998:

> recent strain typing experiments have demonstrated that a new variant of CJD (vCJD), which has so far affected 27 teenagers and young adults in Great Britain and France, is indistinguishable from BSE. These results strongly suggest that vCJD in these individuals was acquired by exposure to BSE prions and has raised concerns that these cases might represent the beginning of a CJD epidemic.[1]

Later, in 1999, the UK Government's Chief Medical Officer, Liam Donaldson, speculated on the possibility of up to 40000 people contracting CJD. By the beginning of 2002, there had been 100 fatalities, and the upper limit on the possible number of deaths had been raised to 50000.

[1] E. Fisher, G. Telling and J. Collinge, 'Prions and the prion disorders', *Mammalian Genome*, **9**, 497–502.

Throughout this emergency, a whole range of veterinary and biomedical professionals was asked for advice. One question that had to be answered quickly was: what was the cause of the BSE epidemic? In Britain there were thousands of BSE cases, in other European countries, dozens. Part of the explanation for the British outbreak lay in the fact that the infectious agent had entered the cow's food chain: there was a problem with the way cattle were fed. The preparation of the high protein, low cost diet needed to fatten cows quickly requires imagination on the part of the feed manufacturers. Discarded fish, bone meal and not-for-sale cuts of animal carcasses have all been considered important elements in the modern herbivore's diet. For decades it has also been standard practice to feed cattle with recycled meat and bone meal obtained from, amongst other animal sources, sheep and cattle. The mix would be processed and heat treated ('rendered') to remove pathogens, but in the late 1970s the process was simplified, perhaps for economic reasons. It is now known that the infectious agent of BSE, the prion, survived the new, cheaper, rendering process. What is less clear is how the prion arrived in the rendering plant in the first place. Some argue that it came from a sheep with the sheep equivalent of BSE, scrapie. Others have suggested that the prion arose by a mutation within one cow. The result at least is clear: the infectious agent was amplified again and again, as the rendering plants, without realising, began allowing prions through into the cattle feed, and recycling them when the cattle had been slaughtered. So the epidemic began. The long incubation period of BSE meant that by the time the Government banned the use of ruminant-derived protein in cattle food (in 1988), the damage was already done: a great part of the British herd had been exposed to possible infection. Moreover, at least 25 000 cases of BSE were diagnosed in cattle born after the ban. Huge quantities of prion-contaminated food, manufactured prior to the ban, remained in stockpiles and in the distribution pipeline, and duly found its way into the cattle trough.[2]

The BSE epidemic was troubling and costly to farmers. A cow that began to show the strange wobbly gait of BSE became a drain of resources,

---

[2] During the 1997 UK general election, the Labour Party (then in opposition) promised that if elected they would hold a full public inquiry. Such inquiries, as the name suggests, hold their deliberations in public, and cross-examine key players (such as politicians and scientists). The Phillips BSE inquiry published full transcripts nightly on its own web site (http://www.bseinquiry.gov.uk/), and revealed wholly disturbing aspects of secrecy in Government, and, especially, a pathological desire to keep information from the public. Interestingly, when the duly elected Labour government faced its own full-scale agricultural crisis (the Foot and Mouth epidemic of 2001), it preferred to hold its subsequent inquiries behiend closed doors, justifying the decision on the grounds of cost and speed. The UK official BSE web site can be found at http://www/defra.gov.uk/animalh/bse/

not a saleable asset. Yet if the costs of vets and slaughter were very unwelcome, they would be as nothing compared with the disaster that would befall the industry if customers stopped buying beef. The nightmare scenario would be if the Government was forced to admit that BSE could transfer to humans, for then it would be very hard to keep people buying beef, when plenty of alternatives were available. Meanwhile, attention was beginning to focus on the obscure human disease of CJD, which was rare, but fatal. The crucial aspect for the current story though is that CJD had already been understood as being the human equivalent of BSE. Some cases are inherited, others caused iatrogenically – that is, by doctors, for example through the use of contaminated surgical instruments, by transplants or even by human growth hormone prepared from human pituitary glands. It had been described as a late-onset disease, occurring mostly in old people. Those people with a knowledge of the BSE epidemic in cattle will have been hoping that BSE and CJD, while the same disease, were nevertheless firmly separated in their affected species: cows for the former, humans for the latter. Thus, when a few young people were reported to have CJD there was widespread publicity. CJD did not normally strike at young people. Could this be evidence of the dangers of beef – that BSE could cross the species divide into humans? The press noted the devastation that the disease causes in a young adult, and in their family. CJD kills by progressively damaging the nervous system; the patient loses physical and mental capacities, eventually needing total nursing care. Accounts from families affected were absolutely shocking; but could it be that these people, barely out of their teens, had been reduced so pitifully simply by eating meat?

For months the Government's position had been simple: 'there is no conceivable risk'. The UK Minister of Agriculture at the time was John Selwyn Gummer. There exists a famous photograph of him attempting to reassure the public by feeding his little daughter Cordelia a beefburger. In this public-relations battle, science got caught up in a way that reveals much about science, as well as the relationship between science and politics. The question 'Is British Beef safe?' was taken by politicians to be a scientific one, and various scientific committees were asked to form a judgement. In their view, there was no evidence that BSE could transfer to people. This typically scientific manner of thinking was construed by the politicians as implying no risk. The Government's Chief Veterinary Officer said in 1990 that 'there is no evidence whatsoever of a risk to human health…The risks to man are remote', a statement that typifies the problems that policy makers have when trying to reassure people by using the concept of risk. The vetinary officer's statement suggested that there was some risk of harm, but that there was no evidence of harm being caused. This kind of reasoning makes some sense to scientists; to others it is completely opaque. An editorial column in one daily newspaper asked in March 1996 'If science cannot help us to explain why and when a hamburger is unsafe, then we are bound to ask what use it is.'

The politicians pushed the 'no risk' part of the equation; the scientists remained stubbornly attached to their statement that 'There is no evidence of any link between BSE and a human equivalent'. When journalists pressed the Government's scientific advisors, it was noticeable that to the question 'Can you say that eating beef is safe?', the answer tended to come 'The risk is tiny.' It came across as two completely different stories, like two tribes unable to agree with each other. It was communication breakdown between politicians and scientists, with the public left to work out for themselves the significance of MacDonald's banning British beef, while Government ministers rejected any possibility of risk. The Government announced loftily in the House of Commons that they were relying on the best scientific advice. Yet when the politicians converted the advice ('there is no evidence of risk') into what they thought was real life language ('beef is safe to eat') the scientists popped up on television and started talking once more about lack of evidence, and remoteness of risk, discussions that had the effect of making consumers more alarmed, not less. Beef prices kept on falling.

In March 1996 it was finally announced that BSE was indeed linked to its human equivalent, CJD. The new cases, occurring amongst young people, were of a different form from the typical CJD. This new variant (vCJD) was evidence that BSE had indeed jumped from cattle to man, just as, earlier, it had jumped from sheep to cattle. Now there was a full-scale health scare, and the time for scientists to be pressed into service for reassurance was gone. The supermarkets cleared beef from their shelves, the European Union banned British beef altogether, and many farmers knew that earnings were about to nosedive.

Yet in one way you would have expected the Government's job to get simpler. Now that the link had been established, it was clear what should be done. The combination of a cattle cull (three million were killed and incinerated) and a ruminant food ban should in time clear the British herd of the disease, and allow a new supervisory regime where every beef cow has an accompanying health record that shows it has had no possibility of coming into contact with BSE. These cows would be safe to eat. Thus, in principle, as the older cows were destroyed, and the younger cows came onto the market, the chance of people contracting vCJD would sink to nil. This knowledge would in turn persuade the public that beef was once more safe to eat, the world ban on beef sales would be lifted, and the fortunes of British farming would be restored. Though the government had a duty towards the beef industry, they nevertheless also had to take notice of any lingering doubts about safety. In late 1997, scientists suggested that there was a slight risk that prions could transfer in cows from the central nervous system into the bone tissue. This advice came at a time when strenuous efforts were being made to reassure foreign markets about British beef. When the UK Government banned the sale of 'beef-on-the-bone' (such as T-bone steak), but declared to the world that British beef was safe, they could safely be accused of mixing their messages. Though beef sales in Britain improved, the prices remained depressed. The European beef ban stayed in place; some education authorities still kept beef off their school menus; above all the scientists were still not saying that they understood exactly how the disease was transmitted to humans, nor what the final death toll from vCJD would be.

The optimistic point of view was that the European beef ban would eventually be lifted, and sales would gradually improve. Skilful marketing – a big sales push – would in time get the beef back into French, Italian and German supermarkets. Those grey-faced scientists, who frustrated the politicians by talking of 'remote risks' would fade from public view. Yet, remarkably, even as the millennium celebrations approached, the beef crisis would not die. What kept it alive, in spite of all the UK regulations put in place at farms and abattoirs, was a simple mismatch between political and scientific thinking. The crisis continued not simply because of important new facts emerging about BSE. More important was the constant media attention to the unstable and unconvincing mix of reassurance and caution that coloured all official comment. The most striking mark of this disparity between politics and science came in the summer of 1999, when the European Commission lifted its ban on British beef, having decided that the UK government had changed agricultural practice sufficiently to eliminate prions from the food chain. The move should have been very good news for farmers, and also for the UK Labour Government. Yet, acting unilaterally, the French Government announced that they were not yet convinced that British beef was safe, and had no intention of importing it: a French food safety committee had advised that they were 'not certain' that the measures instituted in the UK to make beef safe had been effective. The Government in Paris was inclined to be as cautious as possible because of the scandal in the 1980s over HIV (human immunodeficiency virus) and blood transfusions, when French politicians had negligently allowed contaminated HIV-positive blood to be used by the transfusion service, even after a known risk had been established. Some thousands of haemophiliacs became infected. The scandal was such that the law was changed, making it possible for politicians to be sued on matters of public health.

The trauma over the blood scandal strongly influenced the actions of the French authorities. All the French political parties, and all sections of the press, supported the maintenance of the ban. The outrage on the British side of the English Channel led to supermarkets pulling French produce off their shelves, and a press campaign against French beef. Great coverage was given to the story that French farmers had been 'feeding human sewage' to their cattle. On Monday 25 October 1999 a BBC news radio interview with a government scientist probed the safety of putting heat-treated sewage in food. According to the scientist, the treatment would remove the pathogenic microbes. However, the interviewer asked the trap question: 'Can you categorically say there is no risk?' The hesitant admission that risk could not be ruled out was picked up by the tabloid newspapers and converted into the next day's headline 'French beef is unsafe to eat.' By the Wednesday of that week in October opposition politicians were calling for the government to ban French beef, and urging a national boycott of all French goods, from apples to cars. Tony Blair, the UK Prime Minister, loftily refused, saying 'We have science, and the law, on our side.'

This end-of-millennium crisis had a dramatic finale. In banning British beef, the French had forwarded to Brussels, the headquarters of the European Commission, a 600-page dossier explaining their reasoning. It was this document that had to be examined by a European Commission committee consisting of 16 food safety and veterinary scientists drafted from several European countries. If this scientific committee decided that the dossier contained new evidence that British beef was still unsafe, as the French claimed, then the European Commission would support the French action. However, if the scientists found nothing compelling in the document, the Commission would take France to the European Court over illegal practices.

The science committee started its meeting on the morning of Thursday 28 October. By this point the press coverage of the dispute had become so intense and confrontational that it seemed extraordinary to believe that a mere committee of scientists could resolve the problem. What was on show was the idea that the scientific expertise of a group of scientists would avoid all influence by politics or other 'non-scientific' pressure. The British press was intrigued by the possibility of neutral scientists operating under such fevered conditions, and some were frankly disbelieving. There was talk of 'nationalist alliances' between the French and other representatives, and of whether the UK scientists would be 'outnumbered' by the French contingent (it was taken for granted that the French scientists at least could not be neutral, and would sanction the dossier). In this numbers game, a note of interest was that though there were four UK representatives, the chairman, Professor Gerard Pascal, was French. On the morning of the meeting Pascal announced that as some might argue he had a conflict of interest, he was prepared to stand down. He left the room while his colleagues debated this; they decided that as they were 'independent scientists' nationality had nothing to do with it.

It was hoped from the beginning that the committee would reach a unanimous verdict. A divided committee would breathe more life into the lingering perception that British beef was unsafe. The news bulletins that evening announced that the committee had adjourned, and would have to meet on the Friday as well. A lengthy meeting implied a difficult meeting. Journalists outside the meeting barked into their microphones that the scientists inside must be having problems reaching a decision.

Finally, the following afternoon, a press conference was called. The scientific committee had indeed reached a unanimous decision. They declared that there was nothing in the French report that constituted fresh evidence that BSE was not under control in Britain. With that conclusion, the committee had completed its work, and passed the decision over what to do with France over to the Brussels bureaucrats. However, it was noticeable that the committee wanted to make further reassurances about British beef, even if they did not want to commit themselves to saying there was 'no conceivable risk in eating beef'. Intelligently, they found a new form of words: 'There are no grounds for revising the overall conclusions that the safety of UK meat and meat products is comparable to these foods coming from elsewhere in the European Union.' With those words the committee breathed some life into the British beef industry. Yet the relief was short-lived. The French food scientists met on their own account, and found themselves still uncertain over the safety of British beef. Their ambivalence left the French Government once more with a decision to be made, and under the fierce duress of every political party and media organisation in the land, they retained their ban in spite of the certainty of legal action by the European Commission. In the words of the French Prime Minister, Lionel Jospin; 'I'd rather be crucified by the British Press than by French public opinion.'

5.3 Genetically modified crops

BSE was an accident that could have been avoided: different methods of food production would have prevented prions getting into the food chain. As the BSE crisis rumbled on in 1999, with the European Union lifting the ban on British beef on 1 August, and the French unilaterally reimposing it, another food scare was becoming even more prominent. This time, however, the controversy centred on the deliberate, rather than accidental modification of the food we eat. Throughout 1999 there was growing public suspicion of genetically modified (GM) crops. The British press gave great attention to the idea that crops such as soya and maize, engineered to alter their responsiveness to fertilisers and pesticides, were not safe to eat. While safety was the commonest concern in the public commentary on GM foods, doubts about the 'naturalness' of such techniques frequently emerged, often accompanied by the phrase 'Frankenfoods'. Suspicion also centred on the multinational seed companies themselves, who stood to make so much money from the large-scale adoption of the technology. Public concern was strong but diffuse. More focused, and extremely effective, were the mainstream environmental organisations such as Greenpeace and Friends of the Earth, and the wildcat, direct action groups such as Genetix Snowball. They argued that whatever the risks to human health might be, GM crops could lead to 'genetic pollution'. This new form of ecological vice is the transfer of pollen grains, and their engineered cargo of nucleic acid, from GM fields, to fields of conventional, or even organic crops. The integration of modified and conventional genomes was especially excoriated by the Green groups. Such mixing would be irreversible, threaten organic farms, and be in essence unpredictable. Even if the GM crops could be contained, they could still have a damaging effect on the environment. By making the use of pesticides and herbicides so much more efficient, the invertebrates and weeds that survive even the best efforts of conventional intensive agriculture would finally succumb, and so deplete the biodiversity of the countryside.

Another set of arguments about GM crops concerned the multinational companies behind the technology, the most famous of which is Monsanto. Huge sums of money were involved in the development of these new seeds, and it was clear that the companies aimed to recover costs, and make handsome profits, by using some unusual methods. For example, in one scheme run by Monsanto, farmers have to buy not only the company's seeds, but also their herbicides. Attention has also been focused on the 'terminator' gene, engineered into seeds to make the adult plant sterile. The practice of collecting seed from the grown crop, and replanting next season, is thus impossible: more seeds must be bought each year from the company. The strength of feeling, especially in the UK, took the seed companies by surprise. This comment, from an executive at Zeneca Plant Science, was reported in the magazine *New Scientist*: 'Britain could be classed as number two in biotechnology in the world, after the USA. But it's clinging on by its fingertips. These witch hunts are crazy. I feel quite emotional about this at the moment. We still believe in the UK, but God knows why.'

The companies have tried to defend themselves. They point out that in the USA, where GM crops are far more common than in Europe, there has been no evidence of any danger to human or environmental health. They suggest that GM crops can be beneficial to the environment. Maize, for example, is available with a new gene that makes it resistant to the corn-borer pest, so the need for insecticides may fall as such insect resistance is engineered into crops. There are data that insecticide use, especially in the USA and Australia, has fallen with the use of GM cotton. It is argued too that yields will rise, and that this will be especially significant in developing countries, where the human population is still rising rapidly. All these points are in turn disputed by the rebuttal units of the Green lobbyists. Yet the debate is no sterile slanging match. Governments, Greens and seed companies are changing their positions all the time. Monsanto, for example, have scaled down their ambitions in Europe, and have attempted to appease the opposition by announcing that they were dropping plans to develop the terminator gene.

There are some similarities between the BSE crisis and the debate over GM foods. They both involve issues of food safety. They both involve scientists giving advice to politicians about evidence of risk. There are comparisons too in the way that scientists, politicians and the media present different faces of the issue. However, the most important similarity I wish to emphasise at this point is the following: as with BSE, the debate over GM foods has seen politicians trying to reassure the public that GM foods are safe, while the press carries 'bad news' stories. The stories are spawned by the continuing inability of scientists to state categorically that there is no risk whatsoever to this new technology.[3]

The controversy about GM foods had been running in Britain since the late 1980s. The arguments were covered by the press, but did not generate headlines. Though there were strong advocates both for and against, with environmental groups issuing stern warnings about the possible hazards of GM food, while biotechnology companies reiterated the advantages, the issue remained a technical one, not of enormous public concern. No doubt many would have liked things to remain that way, with the safety – or otherwise – of the new crops adjudicated coolly, away from media attention; that was not to be. Early in 1999 the debate erupted as headlines: the foods were linked to Frankenstein. Just as with BSE, the journalists were converting cautious press releases about 'no evidence of risk' into something more sinister. Once released, the story seemed unstoppable. A theme throughout all the media coverage was of the power of the large biotechnology companies, and their alleged hold over governments. Moreover, though scientists might claim the technology to be safe, could they really be considered 'neutral', bearing in mind that GM crops were themselves a deliberate product of science (a situation different from the position that science found itself in over BSE, where the problem had been caused by cost-cutting, not by laboratory workers). It seems that it was the public reaction to industrial and governmental high-handedness, as much as to a genuine concern over gene manipulation, that kept the GM issue burning.

Two small events triggered the uproar. The first was the case of Arpad Pusztai. He alleged that he was 'retired' from his job after telling the British television programme 'World in Action' that the rats he had fed with experimental GM foods had developed abnormalities. His institute, the Rowett Institute, pointed out that to discuss unpublished results on television was against scientific norms.[4] Critics suggested that his work

[3] The debate over GM foods, and the vociferous campaign against them, started as a particularly British phenomenon. Throughout 1999 it was pointed out that since 1996 millions of hectares of American land have been planted with GM soya and maize. When the European anti-GM campaign subsequently started to influence US consumers and farmers this was an unusual example of cross-Atlantic influence travelling west, rather than east.
[4] As is explained further in Chapter 8, scientists have strong views on the way they operate as a community. One is that scientific results should be processed 'in-house' before going on into the

was untrustworthy. Then, a couple of months later, a letter appeared in a quality newspaper, signed by a score of 'eminent' scientists, putting the case for Pusztai. They said he was a world-renowned expert, unfairly dismissed because he said something the establishment did not want to hear. In other words, like Stockwell in Ibsen's play *The Enemy of the People*, Pusztai was a whistle-blower, and had paid the penalty.

This was the story that put GM foods on the front pages. As if dammed up for months, stories rushed into print: the UK Government was in the pocket of Monsanto's lobbyists; supermarkets had not been labelling their foods properly; Britain's wildlife was in danger; science had run out of control yet again; Frankenstein was reborn; the people were being taken for a ride. Throughout the months of headlines, the Government kept to the scientific line: trials of GM crops were needed to assess their danger to the environment. On the question of the safety of such crops, once they were in foods, the Government line was reasonable, but in the light of the BSE debacle, bound to sound weak: 'there is no evidence of any risk' – no doubt many people remembered the same thing being said of BSE in cows before vCJD was announced to be a probable result of eating beef.

The GM food issue can be analysed as follows. Two issues, both perhaps resolvable by scientific investigation, were significant. One was the question of genetic pollution. Accepting the premise that crops might have a harmful effect on the environment, this would have to be investigated. There was the example of the cordon sanitaire: how much space must there be between a GM crop and a natural crop, if cross-pollination is not to contaminate the latter? Similarly, the effect on the crop of wildlife generally can be studied (it was one such study that monitored the possible harmful implications for the Monarch butterfly).

The second issue to be resolved would be the question of whether foods prepared with genetically modified crops could be harmful to human health. It was Dr Pusztai's work hinting that GM foods are unsafe, and his subsequent dismissal, that converted GM foods into a front-page issue. The scientific testing of GM foods is very difficult. Unlike additives or drugs, a 'GM food' is rather bulky. Laboratory animals might not want to eat the food, and certainly will not eat it in the volumes required to do a standard toxicity test. If GM foods carry a risk, is it by virtue of the genes themselves, or their expressed product, or their effect on other genes? To

public domain: in other words, other scientists should get the chance to criticise and evaluate. Pusztai's offence against scientific norms took him outside the community of scientists. For scientists he became a maverick, to the press a martyr.

get these issues refined to the sophistication of drug testing would take years, and seriously delay the full-scale commercialisation of GM foods. From the point of view of the seed companies and governments anxious to help their biotechnology industries, there is a better option. They would prefer scientific trials on environmental impact, creeping commercialisation while the results come in, and a public promise that any GM food will be labelled as such. When the UK government announced such a strategy in November 1999, and a 3-year moratorium on full-scale commercialisation, they said it was time to stand back and reflect on the issue of GM food, and to take some heat out of the issue. No doubt the BSE crisis had reminded the Government of just how unpredictable a food scare can be.

The problem for governments and industry is that their strategy for developing the genetic modification of crops depends on public trust. If people will not buy GM foods, there is no point in planting the crops. Yet persuading the public to relax about GM foods is not easy. After BSE, one strategy is particular is unlikely to be successful: interpreting the scientific claim that 'there is no evidence of any significant risk' as 'GM foods are safe' stands in obvious danger. The reply might well be: 'they said that about BSE'. There are other issues to address as well. At the height of the media frenzy on GM foods, the term 'Frankenfoods' figured prominently. In one newspaper image, Tony Blair, UK Prime Minister, appeared as the actual monster, complete with sickly complexion, a strange hair cut, and a bolt through his neck. The anxiety being exploited was that something unnatural was going on: using science to alter nature itself. The argument has always been dismissed by biologists, for what are simple plant breeding techniques other than genetic manipulation? The Green lobby, on the other hand, attack this argument by pointing out that the gene shifts inherent in genetic engineering are very radical. Genes are being taken from bacteria and spliced into plants. No plant breeding experiment ever managed that. One popular way of describing this is to accuse scientists of 'playing God'. A more measured standpoint, for example as taken by HRH Prince Charles, heir to the British throne and a committed organic farmer himself, is that we simply do not know what the consequences will be of manipulating genes in this way. There is a gulf between the scientific and the lay world view. Biologists, closely familiar with genes, their history and their behaviour, are fairly laid-back about gene technology: they do not see any great risks. The lay person, a consumer of foods and a funder of science through taxes, has not the scientific familiarity with genes, but nonetheless has a sense of their importance and power. To put it another way, a theme running through some critical commentary on the issue is to suggest that genes are too important to be left in the hands of the scientists.

It is interesting to note the Green lobby's reaction to the news that there would be 3 years of scientific trials on GM foods. The reaction was hostile, partly because the UK Government conceded that the crops grown in the trials would be fed to livestock, and thus get into the human food chain. Yet their main attack reminds us of the drama a few months earlier in Paris, when the independence of European scientists over BSE was being questioned in the news. For GM foods are a product of science. Can the results of the trials be trusted? After all, they are being carried out by scientists, representatives of the very profession that has produced this abhorrent technology. By virtue of being generated by scientists, the results cannot be impartial. You might argue, on the contrary, that this is absurd. Only scientists can do scientific experiments; moreover, the whole point of science is that it is necessarily impartial, the simple application of scientific method to the phenomena of nature. Yet whatever your views on the purity of science, it is clear that the worlds of science, money and politics are closely intertwined in the GM food controversy.

Yet the BSE crisis and the GM food controversy show science stumbling in the media, and put on the defensive. GM foods, once simply an innovation in agriculture, have become a model example of how one branch of biology, in this case gene manipulation, can become irretrievably entangled in political and financial issues. It is already true that the success of this technology, judged on its penetration into mainstream agriculture, depends as much on the public attitude to it, as on its ability to deliver the benefits promised by the seed companies. This is a clear example of the success of science depending on social factors, not just on the outcomes of research.[5]

The GM food controversy has been a particularly British problem. In the USA, where millions of hectares are already given over to GM foods, the debate has been more muted. It can be argued that the issues are indeed more important in Britain, with its mix of small fields,

---

[5] It is customary in these debates for a distinction to be drawn between science and technology. The latter is the application of science, whether a bomb or a bridge. The distinction is useful for those who want to maintain that science is independent of social pressures. Applications such as GM crops may wallow in politics and financial dealings, and may even have unethical aspects. According to the argument, however, the science of genetic modification is entirely free of all such side issues. Scientists often make this distinction in order to detach themselves from the ultimate uses of their findings.

still-existing hedgerows, and a remaining sense that agriculture is part of the traditional rural life. In the USA monoculture extends over vast plains; there may be no conventional crops for hundreds of kilometres, and so less of a problem of 'genetic pollution'. Likewise, the wildlife that penetrates British farms, by virtue of the presence of hedgerows and country lanes, is absent from the typical, prairie-style North American agribusiness. Yet there are signs that the arguments are crossing the Atlantic. A 1999 research report suggested that the larvae of the Monarch butterfly are harmed by ingesting pollen from GM crops. The report findings caused some dismay in the USA, where the Monarch is a powerful symbol of American wildlife, and a focus for conservation efforts. The doughty Monarch's great annual migration north from Mexico to Canada is the subject of scores of web sites, and thousands of school students' biology projects. The stories that GM crops are harmful to butterflies were later refuted by other scientific trials, but the initial publicity was a reminder to the seed companies that the safety of American GM crops is a sensitive issue.

5.4 Xenotransplantation

Xenotransplantation is the use of animal organs, tissues or cells in transplant operations between different species, though it is often used in the more specific sense of transplanting from animals to humans. Transplant operations of course are not new: moving kidneys from one human to another has been going on since the early 1960s, and heart transplants became common from the 1970s. The survival rate for transplant recipients is now good enough for the technique to be considered fairly routine. Yet while the medicine of the technique has improved, one problem has remained, or got worse: the waiting lists. In January 1996 the US waiting list for transplants (mostly heart or kidney) topped 44 000. In the UK, the current waiting list is at least 6000. Xenotransplantation might reduce those lists and could impact on more exotic practices too. If xenotransplantation were to become an accepted part of medical practice, we could expect growing interest in the possibility of using pig bone marrow to assist AIDS patients and other immunosuppressed groups, or the transplanting of pig fetal neural tissue to assist people with Parkinson's disease.

Waiting lists are caused by a shortage of suitable donors. Few people who die can actually donate, because the requirements are quite stringent. The heart, lung, kidney or pancreas must be in a healthy state, and removed while the heart is still beating. In practice, donors are patients in intensive care units who are brainstem dead. Such a coma could result from severe trauma such as a motor accident, from cerebral anoxia as a result of drowning or smoke inhalation, or from brain infections such as meningitis. Even when these criteria are met, consent must first be obtained. For a patient to become a donor, irrespective of whether that patient carries a donor card, the family must agree. This they may find hard to do, not because they disapprove of the practice, but because the question comes at a time when decision making is almost impossible: a heart transplant co-ordination team is likely to raise the subject of organ donation at a time of maximum stress, when something very sudden and very awful has enveloped a family.

Some research has been done on attitudes to organ donation. Of 13 000 people who died in intensive care units over a 2-year period during the 1990s, 10% were diagnosed as brainstem dead, making the number of potential donors about 1300 people each year. Of these, some had a medical contraindication to organ donation (cancer, for example); moreover, 30% of families asked about donation refused to give consent. Thus the final figure was about 830 patients, equivalent to 1660 kidneys. In any one year in the UK there are about 5000 patients in end-stage renal failure, urgently in need of a transplant. Thus the waiting list will be long, and many will not survive the wait. The situation is likely to grow more difficult: the population is ageing and roads are getting safer. There were 21 000 deaths on British roads in 1970, and 12 000 in 1990. Ironically, the improved success rate of transplant surgery, due to better post-operative drug regimes, itself makes the waiting list grow: more people view the operation favourably. The situation is similar for heart transplants, though the numbers are smaller: in 1994, some 328 heart transplants were performed in the UK, but 320 remained on the waiting list. This waiting list is itself a slimmed-down version of the much longer list of people who would benefit from a new heart, but who are not considered priority cases.

There have been a number of possible responses to the tragedy of the waiting list. The refusal of families to give consent has been investigated. In general, consent is more likely if the patient is between 15 and 24, or over 65. Families are more likely to agree if they already have planned to cremate the body, and if more than 1 hour elapses between the two tests necessary for confirming brainstem death. Consent is withheld if the relatives do not want surgery to the body, perhaps feeling that the patient has suffered enough. Other common causes of withholding consent are a knowledge that the patient has in the past stated that he or she does not wish donation to take place, or if the relatives are divided over the decision. In sum, the decision over consent is highly particular to the patient, the family, and the circumstances, suggesting that a public campaign on the issue might well be ineffective. It is sometimes suggested, therefore, that the law should be changed, so that relatives no longer have the right to override the wishes of a relative who, before dying, gave consent to organ donation. More radical would be to operate a system of presumed consent. Here, people who did not want organ removal would have to sign a register. Yet the introduction of such a system, with all the inevitable controversy involved, might have the reverse effect to that intended, alienating potential organ donors and their relatives.[6] Of course, another strategy would be to reduce the need for transplant surgery. Public health campaigns on diet and smoking reduce heart disease and, therefore, the need for transplantation. Similarly, the major causes of liver failure in the UK are alcoholism, HIV infection and drug intoxication. Yet such campaigns, even if successful, take a long time. Inevitably, there is interest in developing new surgical procedures, and artificial, or even cloned, tissues may be developed. However, the likely innovation of the next few years will be the procedure of xenotransplantation. Below, I outline research on this procedure, and discuss the ethical and scientific issues involved.

Research since the 1960s has experimented with moving hearts from one species to another. The basic problem is that the massive rejection response necessitates an equally massive, and dangerous, suppression of the immune system. The solution now being investigated is to genetically engineer the donor animal – a pig – so that the identifying antigenic proteins that coat the heart mimic those of humans and so reduce the rejection response to manageable levels. In 1995, a team in Cambridge, UK (Imutran, a biotechnology subsidiary of the multinational drug company Novartis), transplanted transgenic pig hearts into ten monkeys. All were given immunosuppressive drugs; two survived for more than 60 days.

---

[6] In 2000, the Alder Hey Hospital in Liverpool, UK, became notorious for its 'hoarding' of children's organs. It appeared that a pathologist at the hospital had been very systematic in removing organs from the bodies of dead children, but had been much less careful about obtaining the consent of the parents. When parents discovered that they had buried their children incomplete, there was uproar. Second and even third funerals were held for the retrieved body parts, which had nominally been retained for research, but which, it appeared, were simply sitting in bottles with no particular research project planned.

Monkeys given non-transgenic hearts were dead within the hour. Then, 7 years later, in January 2002, the Roslin Institute and the UK based multinational company PPL Therapeutics that created the cloned sheep Dolly, announced the birth of five cloned piglets, each carrying the kind of genetically engineered heart and kidneys that made rejection by humans much less likely. Bred together, these piglets should produce a line of transgenic animals usable for routine transplant operations.

Before considering the scientific issues raised by xenotransplantation, we should consider the ethical issues. The immediately applicable question concerns the ethics of the research into xenotransplantation. Many primates, and many pigs, will be the subject of the continuing experimentation. Through anaesthetics, and good animal husbandry, their suffering will be minimised. Yet they will suffer; for example transgenic piglets will never be fed by their mothers, and they will suffer the undoubted harm of losing their lives. As discussed in Chapter 4, the scientific community generally argues that an experiment harmful to a vertebrate animal is justified by the health and happiness brought to humans. We can see how this so-called 'utilitarian' argument might be brought to bear upon xenotransplantation. Hundreds or thousands of pigs and baboons undergo experimental surgery, with most of them dying. Yet, as a result of these experiments, and those that probe the immune system, many thousands of people are able to receive organs that perform efficiently for many years. These are the people who would otherwise die, or lead a disabled life. Though the ethics of xenotransplantation research per se are issues that should not be ignored, far greater attention is often given to the question of whether xenotransplantation should happen at all: should it be routinely possible for humans to receive animal organs? The fundamental question is clear: should a mammal be bred simply so that it can be dismantled and its parts inserted into humans? Once again, the utilitarian answer is clear. Yes, xenotransplantation is justified because the happiness of the newly healthy human is more important than the suffering of the animal.

Ethical argument is supposed to tell us the difference between right and wrong, even in the face of common sense. It is not the length of waiting lists itself that causes us to investigate xenotransplantation. Much more decisive is the simple fact that we consider the suffering of people more important than that of pigs. However, ethical argument is not simply about principle; detailed knowledge is important too, as we can see when we consider the well-known 'bacon sandwich' argument. This

argument states simply that as the use of pigs in making bacon is toler-
ated by most societies, why should a society wish to put controls on the
use of pigs for providing hearts? This raises the question of whether the
life of a pig destined for bacon is in all essentials the same as that of a pig
destined for the transplant hospital. Those who object to xenotransplantation on the grounds of animal welfare point to some differences between
a 'happy' farmyard, and a laboratory. Pigs on some farms lead fairly unre-
stricted lives, and spend a great deal of time looking after their piglets;
but the pigs that will provide hearts or kidneys live in laboratories, not
on farms. They are transgenic, and too valuable simply to roam around.
They must be prevented from picking up pathogens, and so must live in
a highly restricted, pathogen-free, environment. Nor should transgenic
piglets undergo the risks of birth. They are brought into the world by
Caesarean delivery, and their mother is destroyed. The piglets must then
be hand reared by humans, entirely isolated from pigs or mud or even
straw.

These, in brief, are some of the ethical issues involved. Other ethical
problems concern the question of inessential transplants and the priorities within a health service and a society: should a society spend money
on expensive services that will prolong the longevity of people who suffer from a heart complaint, but are not at risk of death? There is also the
question of whether such a procedure in effect makes an experiment of an
ill person, someone whose condition makes informed consent difficult to
obtain.[7] However, important though these arguments are, it is the problematic science of xenotransplantation, rather than its ethics, that is causing problems and holding up the programme. For xenotransplantation
carries with it a great uncertainty: if you transplant a pig's heart into a human, might anything else get into the human too? Diseases cross species.
Viruses that in pigs are fairly benign might be far more malign, once they
have crossed to humans. It has probably happened before, in the case of
HIV infection: this disease, now endemic with 30 million HIV carriers
worldwide, is believed by some to have come from a primate. So a pig for
transplant must be a pig without viruses.

---

[7] Informed consent is a fundamental aspect of medical research and of surgical procedure. If a person
does not consent to a surgical act, then by law that person is a victim of assault. However, someone
might consent to an operation without fully being aware of the possible consequences. That lack of
awareness could arise for a number of reasons: perhaps the physician did not explain the operation
properly, or perhaps the patient was so desperate to 'try anything' that the possible problems of the
operation were not viewed with sufficient care. The root of the problem is that, as we all know, one
can sign a form voluntarily, without thinking through the issues at stake.

Diseases that cross from animal to human are termed zoonoses. An example is *Salmonella*, a bacterial infection in poultry that in humans can be fatal. The dangers of a zoonosis contracted through xenotransplant surgery are great. Remember, in xenotransplantation, the foreign tissue is placed deep within the person, and is expected to be permanent. To facilitate its acceptance by the body, immunosuppressive drugs are used. If the new organ harbours pathogens, their chances of survival will have been remarkably boosted by the surgery, and by the post-operative drugs. Nor will the viruses necessarily be easily identifiable: it took several years to identify HIV-1 as the pathogenic agent for AIDS (acquired immune deficiency syndrome). As the HIV/AIDS pandemic shows, a virus can spread from person to person for many years before the disease, and its causes, can be recognised.

As with GM foods, we find a scientific development, strongly supported by significant numbers of professionals, languishing because of doubts over safety. Both in the UK and in the USA, the practice of xenotransplantion to humans is banned, although there is now in place in the UK a Xenotransplantation Interior Regulatory Authority (UKXIRA), which was established in 1997, following the conclusion advanced by an earlier advisory group that xenotransplantation to humans could be acceptable, provided that certain criteria were met. The need for new hearts and kidneys grows each year. Will it become possible, soon, to breed pathogen-free, transgenic pigs whose hearts are just as healthy in humans? There is pressure to give a positive answer to this question, and not simply because of concerns about the waiting lists. For example while we tend to see transplant surgery as a last-ditch attempt to save someone's life, biotechnology companies might see things differently. For them, there might be enormous profits in the practice of xenotransplantation. Let us suppose that transplant surgery could become so simple that it could become essentially cosmetic, promising someone, for a price, a better heart, kidney or liver than the one they were born with. If such a practice could become routine, this is where the money would be. The American investment bank Salomon Brothers has predicted that by 2010 there could be half a million organ transplants a year from pigs, ten times the number of ordinary human transplants carried out in 1994. It is estimated that the xenotransplant industry could be worth $6 billion a year – tempting pickings if the scientists, the doctors and the regulatory bodies give the green light. Yet the British experience with BSE, and with GM foods, shows that getting customers to accept a risk over something as important as health, is difficult. Declarations of safety by scientists and politicians have somewhat lost their power. There is something simple, and horrifying, about an animal heart that saves your circulation, but brings as well a viral attack that spreads throughout the population. The idea will be played down by scientists, and the risk minimised; but will the public consider a remote risk an acceptable risk?

6

Making sense of genes

The human genome underlies the fundamental unity of all members of the human family, as well as the recognition of their inherent dignity and diversity. In a symbolic sense, it is the heritage of humanity.

Universal Declaration on the Human Genome and Human Rights, 1997, Article 1.[1]

6.1 An introduction to the Human Genome Project

This chapter is about the Human Genome Project (or HGP), biology's best funded, and best publicised, research effort. It was launched in 1990, and reached a conclusion in 2001, when provisional results were announced. During the intervening 11-year period the science of human genetics was seldom out of the headlines. The high profile of genetics relied on a stream of announcements linking genes to diseases, or even to behavioural characteristics. Not only was it commonplace to hear discussions about the gene for breast cancer, cystic fibrosis or Huntington's disease, there were also claims about the existence of genes for homosexuality, alcoholism, intelligence and aggression.

The Human Genome Project has two fundamental aims: the first is to make a map of every human chromosome, with each gene properly positioned and identified; the second task is to sequence the genome, that is to list in correct order the three billion base pairs that form the linear nucleotide structure of the chromosomes. The technical demands of those tasks do not form the focus of this chapter. Instead I am interested in exploring the way that the project has become so significant a feature of the biological landscape. Even more than that, however, this chapter looks at

[1] http://www/unesco.org/human.rights/hrbc.htm.

some of the issues surrounding the use of genetic information in modern society. The Human Genome Project is not the first exercise in gene mapping, or in DNA sequencing. Mapping began in the 1920s, when a few genes of the fruit fly *Drosophila* were mapped to positions on individual chromosomes. The first techniques for sequencing DNA were developed in the 1970s, and were in common use in the 1980s, but to take these ideas and apply them to our own genome, the Project had to become a high-profile, famous, biological venture. It needed unusually high levels of public funding. The vastness of the task necessitated an extraordinary amount of collaboration between teams of scientists, and as the Project advanced, the technical expertise in automation and bioinformatics developed too, and made it more likely that its goals could be reached. Another striking feature of the Project has been the ambitious descriptions of the potential outcomes. For along with the physical map, molecular biologists would isolate genes and study their mutations. The lengthening list of isolated genes would certainly include many involved in disease: these could be used for inventing tests for the disease concerned, and for studying the disease itself. That makes the Project a piece of medical research, and not simply biological research. It is this medical aspect that has generated the highest hopes. As the UK Prime Minister Tony Blair said in 2001: 'Today we are witnessing a revolution in medical science whose impact can far surpass the discovery of antibiotics.'

The 1990s saw an immense amount of commentary and controversy concerning the Project. The simplest controversy was over the expense, and its possible distorting effect on the biological sciences: if human genetics becomes a priority for research funds, does this mean that other areas of biology will be left unsupported? Thus, in 1990, as the project was being launched and its US federal budget finalised, a group of microbiologists from Harvard Medical School wrote to the journal *Science.* They suggested that the Project was getting too big a slice of the biomedical research pie – having estimated the size of that slice as 20% – and questioned how this would affect the ambitions of other areas. Inevitably too there were political rows over who should be in charge, and scientific altercations about the methods and targets.[2] From the point of view of this

[2] A detailed summary of the development of both the politics and the techniques of the Human Genome Project can be found in Robert Cook-Deegan's *Gene Wars: Science, Politics and the Human Genome* (New York: W.W. Norton 1995). For a fuller account of arguments over the ethical implications, see *The Code of Codes: Scientific and Social Issues in the Human Genome Project,* edited by Daniel Kevles and Leroy Hood (Cambridge, Mass.: Harvard University Press, 1992). Inevitably most academic texts give only a very partial sense of the pain and fear of inherited disease, and a compelling corrective is *The Troubled*

chapter, the most important cluster of problems arose from questions over the meaning and use of genetic knowledge. The fact that the Project actively sought out genes, and then publicised them, inevitably focused attention on the old question: how important are genes in shaping human lives? It is a question that people approach with different motives. Consider the likely interest of families with genetic disease, compared with any interest in personal data exhibited by agencies (such as the police, or insurance companies). Amongst the academics, while philosophers might analyse the meaning of the term gene, and biologists might research the role of genes in the mechanisms of development, biomedical researchers might look for new therapies based on new knowledge – and might also seek to make their fortunes.

This chapter focuses on three areas. First I investigate what kind of biology is represented by the Human Genome Project. Earlier in the book, I explored the point of view that might regard, and even condemn, the Project as nothing more than a gigantic technical exercise in listing nucleotides, and so an exercise both reductionist and meaningless. This accusation would certainly be reasonable if the Project scientists did indeed insist that the most important piece of knowledge a biomedical researcher can have is a list of nucleotides. Of course, the Project has involved more than listing nucleotides: it has also mapped – located the position of – genes, but how far does that knowledge go in predicting the behaviour and health of human beings, or in guiding with irreproachable success the investment policies of pharmaceutical companies? If your only guide to the new genetics was the stream of media stories about gene discoveries, you would certainly imagine that the Project proved the paramount importance of genes in human health and behaviour.

My second theme concerns the ethical implications of the Human Genome Project. By this I mean the implications that genetic advances can have in generating new and possibly unwelcome choices. For individuals these choices (and decisions) might be whether to have a genetic test, and whether to tell other family members the results. For a university department; the choices might be whether to attempt to patent, and make money from, a technique involving a human gene. There will be decisions too for commercial agencies, such as insurance companies: should they insist

Helix: Social and Psychological Implications of the New Human Genetics, edited by Theresa Marteau and Martin Richards (Cambridge: Cambridge University Press, 1996). This interesting collection devotes its first 60 pages to personal stories of patients and families, and gives insight into the feelings associated with genetic testing.

on being told the results of all genetic tests? Finally, the police, state governments and even the United Nations will have to make decisions about when to exploit genetic knowledge, and when to use the law to protect citizens.

My third theme simply involves a 'thought-experiment:' how will visits to the doctor be affected by genetic testing. So-called 'predictive genetics' will allow assessments of future health. The tests might be done early in life, on the fetus for example, or even on a gamete. The information your doctor has about you might almost be a prison, because that information is a picture of your future. You and your doctor, therefore, will not simply discuss how you feel now; you will be discussing genetic information that gives clues over how you will feel in 5, 10, 15 or even 30 years time.

6.2 The philosophy of the Human Genome Project

The goals of the Project were set out on its launch in 1990. The estimated 80000 genes that make up the human genome would be 'physically' mapped;[3] that is, the precise position of genes would be discovered not by the usual allele recombination studies, but by looking at the DNA itself. At the same time, laboratories across the world would use increasingly sophisticated and automated machinery to list the three billion base pairs that constitute the DNA of a human being. Both tasks have been aided by the development of databases and software that analyse the information generated, and an important consequence of the Project has been the sudden expansion of bioinformatics, the application of computing power to biology. Another aspect of the work has been deliberately reflective: unusually for a science project, philosophers and sociologists have also benefited. Money has been put aside to encourage scholarship and debate about the ethical, legal and social implications of genome research. This is because politicians and scientists understood from the beginning that the promises made about the value of the project, and its expense, would guarantee headlines. Moreover, irrespective of how the media ran genetics stories, genuinely new and difficult ethical issues might arise and need proper analysis, and it would be important to have these studies

---

[3] This was a common estimate of the number of genes at the start of the Project, but by the time the first draft was published, 11 years later, the best estimate had fallen to less than 30000. This reduction prompted a fair number of headlines, probably because the new estimate was not far from the number agreed for the mouse genome. The irresistible popular question was 'in that case, what makes a man different from a mouse?' One newspaper referred to the human genome as containing 'little more genetic information than mice, and scarcely twice as much as tiny fruit flies'.

going on alongside the project, rather than afterwards. Thus, when James Watson was appointed head of the office of genome research at the US National Institutes of Health (NIH), he declared that a proportion of the budget would be set aside for study of the social implications.

The Project started formally only in 1990, but had a long gestation period. Throughout the 1980s, pressure had been building towards an 'attack' on the human genome. This particular decade had seen a fast-developing expertise in techniques for manipulating DNA. Smaller genomes had been studied; now the talk was of doing systematic work on the biggest and most important genome of all, that of *Homo sapiens*. The idea at its simplest was to sequence every single base that makes up human DNA. This was to be complemented by mapping to a chromosomal location every single strip of DNA that could be identified as a gene. Gathering the data would be expensive, time consuming, and, some suspected, mind-numbingly tedious. PhD students and robots would do the work, with automation becoming more and more important. As time went by the dollar-per-base cost would fall, and the speed of the work would increase.

Such work is expensive to the tax payer, and it must be justified. In the public relations exercises that surround the Project, the emphasis is usually on its medical implications. Knowledge of genes and of their mutations can link physiological processes to genes. It allows a much richer filling out of the causal pathways in organisms. This does not necessarily mean that the idea of an organism as a 'lumbering robot, controlled by genes' will prevail. On the contrary, it could emphasise a different view entirely. The Project could accentuate an interest in the way that genes and proteins and the environment interact, and illuminate the interplay between the higher-level and lower-level explanations discussed earlier in this book, in Chapter 2. Still, it must be admitted that the Project is often interpreted as some kind of a triumph of reductionist biology. The language surrounding it has often been grandiose, as though the sequencing machines were generating not only long lists, but also 'the secret of life itself'. The rhetoric suggests that genetic knowledge, even knowledge of nucleotides, is the most important information that a biologist can have in understanding organisms, or a physician in understanding disease. Some of the least publicised criticisms of the Project have come from biologists, concerned that the proponents of the Project overemphasise the importance of genes in determining behaviour or illness, and are thus in danger of reducing the importance of social programmes for combating diverse problems. For example alcoholism might have a genetic link, but for the time being the most effective routes towards reduction of the incidence of alcoholism are likely to be through a knowledge of society rather than genes.[4]

In spite of the publicity given to the Project, it is probably true that scientists have become more modest about the significance of the new genetic knowledge. When the draft was announced in February 2001, the scientific commentary discussed the medical cures that would 'in time' become available. The most immediate result was the launching of a lot more work in genomics (the study of the expression of genes) and the related proteomics (the study of proteins encoded by the genes). These successor schemes show that, even if the initial ambitions of the project were very reductionist, the ensuing work will explore interactions at higher levels – at the level of proteins, for example. Little is known about how proteins interact with each other and with the environment and it is very likely that future research into organismic development or into genetic factors in human disease will draw on a wide range of biological, physical and environmental phenomena. An optimist might conclude that the draft of the human genome is very obviously not a description of human life, and that attempts to put the genetic knowledge to practical use will revivify the traditional breadth of biology and bring together reductionist and non-reductionist ways of working.

However, before ending this brief discussion of the philosophical background to the Human Genome Project, I will outline a startling thought-experiment that expresses the idea that a complete knowledge of genes will give a complete knowledge of the organism; the thought-experiment is just that – it would be very likely impossible even in principle to execute it. In considering it, however, you might be able to consider your own reaction to it, in particular how well it agrees or conflicts with your own views on genes. The basic premise of the experiment is the idea that genes are more than an internal description of the organism. They are also the instructions for turning a description into a body. The organism, in other words, can be seen as a computer, executing its own programme. According to this metaphor, we should in principle be able to garner the genetic

---

[4] This example and many others can be found in *Lifelines: Biology, Freedom, Determinism*, the excellent work of criticism by the biochemist Steven Rose (Harmondsworth: The Penguin Press/Allen Lane, 1997). Other texts that question the new genetics are *Not in Our Genes: Biology, Ideology and Human Nature*, edited by Steven Rose, Richard Lewontin and Leo Kamin (Harmondsworth: Penguin, 1984) and *It Ain't Necessarily So: The Dream of the Human Genome and Other Confusions* by Richard Lewontin (London: New York Review of Books/Granta, 2000).

description of the organism, and also the rules by which the organism is made, simply by studying the genome. With enough computing power, you could replicate the description and the instructions, and model the complete organism: the creature would appear on your screen. You can imagine a college practical that involved downloading from the Internet the complete sequence of an organism. You have no idea of what the organism might be, but your exercise is to use the data to model the organism, building up a description of its form, its physiology, its sexual behaviour and its life cycle. Highest marks will go to the student whose organism most closely fits the actual one used to derive the sequence. No such college exercise exists now, but could it in the future?

You will recognise in this thought-experiment its key assumption: that an organism is captured by its genome. Certainly the genome is needed to make an organism, but does the genome contain all the information necessary for making that organism? An analogy would be the musician who reads a score, and starts humming the music. Only a non-musician would imagine that the notes on the page absolutely determine the sounds the musician makes. The truth is that the musician is not only the means by which the score becomes music, he is also an interpreter, an active agent in the making of that music. He is also, of course, 'separate' from the music, yet as a musician he could not exist without it. The same situation, it could be argued, exists with cells and genes. Cellular chemistry is needed for the expression of genes, so genes depend on cells, and yet we know that there could be no cells without genes. Partly because of the Human Genome Project, it has become even more obvious that the metaphor of a gene as a word, passively waiting to be read, is becoming too inaccurate. Genes – especially human genes – do not always have fixed meanings: parts of some genes might be spliced together with parts of others, even from a different chromosome. This concept of 'alternative splicing' makes the idea of genes as fixed particles less useful. Instead, other metaphors for describing genes (however crudely) are emerging: plastic, sensitive, organic. It is interesting to think that as genetics proceeds, our metaphors for describing genes will keep on changing too.

Though the Project raises interesting questions about genes and about reductionism, its importance has tended to be described as medical. A debate over whether the Project is best seen as biological or as medical would be long and complex, and beyond the reach of this chapter. I shall make a simplification here, and assume that the Project has become a branch of medical research, brought into being because it can allow analysis of the genetic components of human disease. The importance of gene mapping can then be considered in this way: if we have a map of the human genome, and have isolated many of the genes that cause or are associated with disease, then we can begin to test people for their susceptibility to genetically caused disease. That way we may be able to predict whether someone becomes ill through possession of a mutant form of a gene. The supporters of the Project suggest that we might be able to find the right cure, or to suggest preventive action. Perhaps we will be able to repair the gene, or replace it, or knock it out. Moreover, we can move beyond those diseases caused by single-gene defects. With the data of the Project we should be able to increase decisively our understanding of all those diseases that have a genetic component (and we should remember that fully two out of three of us will die from a condition that has some genetic basis). The task will be complicated, and long term, and expensive. Nevertheless, with the nucleotide sequence, and the gene map, and a lot of extra research, we will be able to tell a person the diseases he is likely to fall victim to. Project supporters have also pointed out the benefits of offering such a person either a cure, when once it had seemed impossible, or, if a cure proves too difficult for a few decades, the possibility of minimising the effects of the gene – for example by following a particular diet, or health regime.

6.3 The history of the Human Genome Project

The first mapping of a human gene took place in 1911. The gene was the one associated with colour blindness. Two observations made this possible. Firstly, colour blindness is much more common in males than in females: it is sex linked. Secondly, the chromosomes apparently responsible for sex differentiation are remarkably different in appearance, with the X chromosome distinguished from the Y by its larger size. Mapping, therefore, relied on a knowledge of patterns of inheritance, and on visible differences in chromosomes. Progress was bound to be slow because in the first half of the twentieth century it was difficult to distinguish the different somatic (non-sex) chromosomes. A gene could be mapped when a particular disease running through a family showed up only in individual family members who possessed a particular physical change in one of the chromosomes. This kind of gene mapping, dependent on a knowledge of family history and on gross chromosomal abnormality, could not identify more than a handful of genes, and managed only to map them on a particular chromosome, not to a specific position on that chromosome.

Before the Project began in 1990, only a few human genes were mapped. An estimated 80 000 genes lay unknown, either by function or by position. Yet the techniques for mapping genes more precisely had been developing since the 1930s. The majority of gene mapping took place in organisms where experimentation was possible; an example is the use of radiation to produce mutations. From the 1960s, molecular biologists were getting close to the direct manipulation of genes. The discovery of restriction enzymes made it possible to cut and edit DNA. Bacteria have been useful too, because they can be cultured by the million and their prokaryotic genome is easy to manipulate. The first examples of 'genetic engineering' emerged in the 1970s. One well-known medical application was the insertion of the human gene for insulin into bacteria, which then began to produce human insulin themselves.

Nucleotide sequencing techniques were invented in the 1970s. The first genome sequenced was the Phi-X virus, with 5 386 base pairs. Next was the DNA of the human mitochondrion (16 000 base pairs). In 1984 the sequence of the Epstein-Barr virus was announced (172 000 base pairs). The largest sequence published during the 1980s was that of human cytomegalovirus (229 000 base pairs). However, viruses are not even organisms: could the genome of a multicellular animal be mapped and sequenced, and used (for instance) to investigate development? The information would be useful, as well as interesting. Such is the conservatism of evolution that the genes of a small invertebrate might well be found to have a similar role in human beings. In fact, bearing in mind the fact that two-thirds of all human disease has a genetic component, small-scale genome sequencing projects would have a more obvious implication for medical research if they could be extended to a multicellular (metazoan) organism, a worm perhaps, and then later, once the relevant techniques had been developed, to a human. The metazoan chosen was the nematode *Caenorhabditis elegans.* This was suitable because of its small number of cells, each of them visible under a microscope. Because of this visibility, the development and movement of each cell can be tracked, establishing a pedigree from single cell zygote to adult. By mapping the genes of the nematode, and sequencing its base pairs, the genetic control of development could be studied. The information was interesting both medically and biologically, for as worm genome specialists like to say, a human is 'just a worm scaled-up'. The techniques and automation developed in working on the nematode were also important in giving scientists a sense of the plausibility of setting up a genome project on a much bigger and more complex metazoan: the human being. From 1985 therefore, a large but cohesive community of nematode biologists began to develop. Their research, and the developments in technology that accrued as the worm work continued, were to provide valuable insights into how actually to organise a Human Genome Project.

The nematode research itself took place on several fronts. One laboratory made 20000 electron micrographs of the nervous system, and watched the behaviour of its constituent cells – all 302 of them. Another laboratory traced the origin of every one of the 959 somatic (non-sex) cells. Genome scientists estimated that nematode DNA runs to about 100 million base pairs and started mapping an estimated 700 genes onto six chromosomes. By 1988, the relative position of each gene had been found, and the next stage began: the sequencing of the DNA. Sequencing techniques were by now automated, and fast enough to suggest that the goal of listing every base pair could be achieved fairly rapidly. The progress of the laboratories studying *Caenorhabditis elegans* confirmed the view that the work on nematodes could be extended to *Homo sapiens*. More people and much more money would be needed, but with the nematode work a success, mapping the human genome seemed less a fantasy, and more a technical problem.

6.4 Genetic testing: the case of Huntington's disease

Huntington's disease is one of the few thousand diseases linked to a single, defective gene. The mutant allele is dominant: you can inherit the disease from one parent only. A particular aspect of this inherited disease is that if someone has the gene, the symptoms will show only late in life. Thus, you can develop into an adult, be completely healthy, and bring up a family, before any symptoms occur. In that time the gene might have been passed down one, or even two generations. When the disease does finally strike, the symptoms are varied, but grow more severe. The Huntington's disease patient succumbs to depression, panic, loss of memory, extreme irritability, apathy. These symptoms, which make life extremely difficult both for the sufferer and the family, are accompanied by more physical effects, including paralysis. Eventually, total care is needed; the disease is invariably fatal. If there is a history of Huntington's disease in a family, members will surely ask themselves: will I live to take my pension, or will I inevitably die early, unable to hold a fork or remember friends' names? The issue might be discussed openly, or it might be a 'secret', with brothers and uncles and aunts looking at each other for signs of erratic behaviour. There is no cure for the disease, nor any treatment. Yet, as the disease is so plainly a genetic one, the one hope has been that knowledge of the gene will lead to knowledge of a cure.

The actual gene for Huntington's disease was located in 1993. As is often the case with gene mapping projects, this knowledge cannot be linked to the pathology. Yet it is clear what is wrong with the gene, for it is faulty in a very simple way: in Huntington's disease, the mutation involves a stretch of DNA where the codon CAG is repeated. In the normal gene, the codon repeats between 11 and 34 times. In the mutant form, the codon is repeated between 42 and 66 times. The greater the number of repeats, the worse the illness. A person who has Huntington's disease in the family can be tested for this genetic fault. Thus, families affected by Huntington's can have a genetic knowledge of themselves, but cannot use that knowledge to do anything about the disease.[5]

There is no such thing as a 'typical genetic disease'. However, Huntington's disease, and the experience of affected families, tells us something about what the new genetics can, and cannot, do. In particular, the case of Huntington's teaches us a basic, salutary fact about the current utility of the information coming from the Human Genome Project: you can determine a gene, sequence its base pairs and identify the mutation, and still have no idea about how the gene causes the disease. Huntington's is an early example of a phenomenon that will become more and more common: a hi-tech genetic diagnosis is available, but no cure. The huge research effort in the 1980s and 1990s that discovered so much about the Huntington's gene, has had absolutely no impact on the development of the disease in patients.

Yet knowledge of possession of the gene, however little it has affected the abilities of the doctors, does affect the life of the family with a history of Huntington's diseae. For with the mapping of the gene has come the development of the test. It is possible to be tested, and to get a clear positive or negative result. Instead of waiting, perhaps for 50 years, to find out whether or not the disease will strike, someone who has seen Huntington's disease in the family can put themselves forward to be assessed, and can be declared free of the gene – or not. The choice, therefore,

[5] This is the paradox of the new genetics. Actually the situation is not new. Until the end of the nineteenth century, doctors had a rather restricted ability to intervene medically, yet their knowledge of anatomy was likely to be excellent. Until sepsis and anaesthesia and drugs were developed, there was little a doctor could do – except listen to his patient.

is not over types of treatment, for none is available. The choice is over whether you want to be tested, or not. It is a new kind of decision, and a difficult one. Do you want to know more about your future? Who will you tell that you intend to have a test? What will you do with the results?

You might imagine that, at least in the case of Huntington's disease, it is always best to know. In fact, since testing has become available, it has become clear that members of affected families often do not want to know. Having a test, and learning the result, may have unforeseen effects. A person may have spent years thinking that he may get the disease, and then finds he is free of the gene (strictly speaking, the mutant allele). Perhaps the same test, applied to his brother, gives the opposite, result. How will the relationship between the two brothers be affected? Or take the case of the young woman in an affected family, with two apparently healthy parents. She has the test, and finds she has the gene. That must mean that at least one of the parents must by definition have the gene. Perhaps those parents a few years ago, faced with the fact of the test, opted to avoid it after months of thought. We can see that profound issues about one's idea of a future, and about one's family, are raised. It is not surprising that people offered a gene test often decline to have it. Research shows that those who have lived with the possibility of a late-onset genetic disease, but are tested and found negative, have the most conflicting feelings about their 'freedom'. Some have found themselves free of the gene, but on the therapist's couch, suffering from depression. This, then, is a family affair.

The medical profession also has to face issues over genetic testing. For example what kind of patient preparation should precede a genetic test? Is it a doctor's job to make sure that a person wanting a test is fully aware of the implications? If the advice is going to be time-consuming, how should this be paid for? Undoubtedly the demand for genetic counsellors will grow, and become a new expense. The genetic tests themselves may be expensive, because they must be manufactured by private companies. Should such companies be allowed to advertise their tests directly? Or would this lead to doctors overprescribing such tests, in the same way as they might overprescribe antibiotics?

There is one certainty with the gene associated with Huntington's disease: if you have the gene, you will certainly get the disease. This is referred to as 100% penetration, and is unusual. More often, when a disease is linked to a gene, simple possession of the gene does not imply illness. We see this in the basic genetics of cystic fibrosis, a disease caused by a recessive mutation; people heterozygous for the cystic fibrosis allele are carriers, but show no sign of the disease. Yet very often human genetics does not follow the simple rules of Mendelian ratios. For example a gene test for diabetes is not likely to give a result that is simply positive or negative. The genetics are too complex, and are intertwined with environmental causes. Moreover, a test might not be able to pick up every kind of mutation occurring in a gene. In the case of cystic fibrosis, when the gene (the CF gene) was isolated there was strong hope both for a test and a cure; cystic fibrosis is one of the commonest serious single-gene diseases in Europe and North America, with one in 20–25 people thought to be heterozygous for the gene. After the gene isolation was announced in 1989, a UK opinion poll showed that 80% of those who had heard of cystic fibrosis wanted to know if they were carriers. Yet however impressive the scientific breakthrough might be, there are problems in testing for the CF gene, for example there are many different mutations affecting the gene, not all of them detected by the test. No doubt different CF mutations have an effect on the incidence or severity of the disease. With time, some of these uncertainties may be resolved, but the case highlights a specific problem with tests – their reliability, and the expectations that people may have of them. Making sense of test results is likely to be time-consuming and a challenge for the communication skills of medical professionals.

Even if the accuracy of tests for single-gene defects improves, the most usual result of a genetic test will not be a 'yes' or a 'no', but a risk assessment – that is, a calculation of the chance that you will suffer from the disease in the future. This is true of the most important Western diseases such as heart disease or cancer, where many genes may be involved, and there are environmental influences at work too. Once again we find uncertainty being an unavoidable fundamental outcome of research.

6.5 Testing for the 'gene' for breast cancer

With Huntington's disease we saw that affected family members are often ambivalent about taking a gene test. The decision to take a test may be even more difficult if the information given is in the form of a probability. The problem is greater still when the test produces a risk assessment for a disease that medicine cannot prevent. Breast cancer, and its link to the BRCA1 and BRCA2 genes, provides us with an important example of just such a disease.

Sometimes, breast cancer runs in the family. In such a family, though some women remain free of the disease, its incidence is unusually high.

Coping with the disease, or with the thought that one might die from it, can then become a family 'issue'; as with Huntington's disease, some families are open and forthright about the matter, others less so. Yet not all breast cancer is hereditary; in fact much of it is not. Breast cancer as a whole is so common that all women over the age of about 55 are checked every so often by having mammograms, (X-rays of the breast). Such population-wide testing is called screening; the expense may be worth it because the cancer is common and, caught early, can respond well to treatment.

The widespread incidence of breast cancer, and the fact that some of it is familial, triggered in the early 1990s an intense 'gene hunt'. Eight teams hunted for the gene; the winner was a team in Utah University, USA, who used the meticulous family records kept by the Mormons in Salt Lake City. The university team, reborn as Myriad Genetics Inc., promptly sought to market a test that would allow a woman to find out whether or not she has the so-called BRCA1, or breast cancer gene (actually a mutation in a gene whose function is not well understood). There was immediate controversy, not simply because of the rapid commercialisation of the discovery but because it was not clear that a test for the BRCA1 gene would actually help anyone apart from the shareholders of the biotechnology company concerned. The problem lies in the fact that breast cancer occurs even in women without the gene; conversely, women with the gene do not definitely get cancer. With enough testing in an affected family, and with the right kind of expert advice, it is possible to give a clear indication of risk. However, when the interpretation of a test result requires hours of expert counselling, together with parallel testing of other members of a family, questions of finance are bound to arise. The idea of the simple test being available, without such back-up, is controversial. On the one hand there is an argument that a women has 'the right to know'; on the other hand there is doubt that the test result constitutes useful knowledge, unless supplemented by advice or counselling. The suspicion is that the test does not provide enough information to guide the health decisions of the patient.

Unsurprisingly, the two genes currently known to be involved in breast cancer, BRCA1 and BRCA2, are under intense scrutiny. The BRCA1 gene is a large one, and can carry many different mutations; the gene test for it may pick up some mutations, but not others, and the significance of each mutation in causing the disease is not known. It is not even known why a mutated BRCA1 gene should be associated with breast cancer, though research has shown that the normal role of the gene is for DNA repair. What is known is that BRCA1 and BRCA2 predispose a women to breast (and ovarian) cancer. What remains unknown is the mechanism of causation, and why possession of the gene does not inevitably lead to cancer. If breast cancer was a minor disease, and its prevention or cure a trivial procedure, the inadequacies of the gene test would be less important. The problem is that a woman diagnosed as 'at risk' has no simple prevention available: mastectomy, removal of the breast, reliably closes the risk, but is an extremely radical step hardly definable as 'prevention'. Ultimately, it is the seriousness of the disease, and the absence of a non-invasive prevention strategy, that makes the BRCA1 test so controversial.

A disease as common as breast cancer, with so complex a genetic and environmental background, and with so varied a survival rate, is bound to generate a strong desire for knowledge, not only amongst doctors, but amongst patient groups too. These patient groups are extremely well informed, and reveal the falsehood behind the prejudice that the lay public knows no science. A patient with a chronic disease is likely to be very up to date on the relevant pharmacology, perhaps more so than a general doctor. This democratisation of scientific expertise occurs through meetings, through journalism, and, especially, through the Internet. A search for web sites about breast cancer will list more than a million, a good number of them devoted to giving women access to the knowledge they need in order to make informed choices about treatment. However, breast cancer is too complex and poorly understood for a web site to provide simple answers; the available facts about the incidence and treatment are overwhelmingly statistical.[6] Yet the cancer patient, or someone considering a gene test, wants to find out about their own body, not that of an average person. Like all sciences, genetics produces great amounts of factual information, and great amounts of speculation, with the distinction often unclear. The genetic output discusses molecular events and statistics concerning patients. Morbidity (amount of illness in the community), death rates and recovery rates are useful numbers for physicians and politicians, but do not easily translate into the experience of a patient. Gene tests may give a person more information about their body,

---

[6] The basic statistics run like this. Breast cancer is a disease that one in ten women develop. Eighty per cent of women have no family history of the disease; most breast cancer has no genetic link at all. In the USA, 180 000 cases of breast cancer are diagnosed each year, a huge number. Yet treatment is very often successful. Some 97% of those whose cancer is detected early will respond to treatment and return to full health (this demonstrates the importance of breast screening programmes), but women who inherit a single mutated copy of either the BRCA1 or BRCA2 genes have a 40% increased chance of early-onset cancer.

but some uncertainty will still remain. The truth is that a disease with a complex genetic component generates worries that cannot be banished simply by a gene test: the gene test, performed in the right way, and with the appropriate counselling, will offer valuable information but will not remove anxiety. Instead, the anxieties will tend to be transferred; instead of worrying about whether you are at all likely to get a disease, you may debate instead the statistics about the success rate and nature of available treatment. Information that you are predisposed to a disease will trigger other questions, perhaps about lifestyle, self-esteem, or family relations.

6.6 The Future of the Human Genome Project

We have discussed in this chapter a development in biology (nucleotide sequencing and gene mapping) that grew into a giant medical project. No doubt we view the Human Genome Project as medical because of the vague but loudly proclaimed slogans about a coming 'medical revolution'. Usually, when this bright future is discussed, gene therapy is mentioned, and no doubt there will be cases where cures will become possible by knocking out faulty genes, or by simply inserting new ones. This route is being followed by those seeking a cure for cystic fibrosis. Here, the fact that cystic fibrosis affects lung tissue makes it a good test bed for gene therapy. Surface epithelial cells of the lung are easier to reach than cells inside bones, for example. Moreover, cystic fibrosis is a single-gene disorder: in principle, only one kind of gene need be engineered. In fact, there are enormous problems with gene therapy, which are related to the efficiency and safety of the vectors used to transfer the genes into the patient, and the fact that the majority of genetically associated diseases cannot be linked to one gene or even to a few – many are involved. Moreover, getting engineered genes inserted and functioning in the right place is difficult. It is not likely that a medical revolution is imminent. Instead, decades of research will gradually find ways of using genetic information to develop new cures and more effective explanations of disease and to improve the quality of drug therapy.

In February 2001, *Nature* magazine devoted an entire issue to publishing a draft map of the human genome. For the first time, formal scientific papers laid out an overview of the make-up of the 23 chromosomes. It was an important event, and generated a large amount of press coverage. As usual, the comment speculated on the new medical cures that would certainly follow. Yet at the press conference that accompanied the release, the scientist most closely associated with British genome research, Sir John Sulston, was very circumspect. He noted that the published information was just a start, and said 'It requires tens of thousands, hundreds of thousands, of good minds looking at it in an unfettered way, doing experiments, to really make the contribution that the Human Genome Project information is capable of.'

The *Nature* publication included a revised estimate of the number of human genes, reducing it from 80 000 to about 30 000. The estimate emphasises, if emphasis is needed, how crude an error it is to think of genes as read-only strips of DNA, each one dedicated to producing a single trait. The estimated gene number for mice is 22 000, suggesting that the greater complexity of a human is not due simply to increased gene number; indeed, a plant, *Arabadopsis*, has 26 000 genes. Instead, genes, and parts of genes, must have some flexibility in how they are used, being edited and expressed in a variety of ways, rather than in one way. Understanding these activities, and making use of them, will be a task far removed from the automated business of sequencing DNA. Undoubtedly, these new areas of research, still hardly sketched out, will involve explanations occurring at every level: molecule, cell, organism and environment. As I suggested at the beginning of this chapter, we can expect that the reductionist phase of genome research will fade, to be replaced by a much wider-ranging series of questions. Some of the genes mapped, it appears, have jumped into our genome from bacteria, and subsequently been found to be useful, so they have been allowed to persist. Proteomics, the study of protein expression and control, will be the next phase of research. The interest of the task was hinted at by the authors of the 2001 *Nature* sequencing paper by the International Human Genome Sequencing Consortium,[7] when they echoed the original paper by James Watson and Francis Crick announcing in 1953 a structure of DNA.[8] In that paper there exists a famous piece of understatement: 'It has not escaped our notice that the specific pairing we have postulated immediately suggests a possible copying mechanism for the genetic material.' The 2001 paper, unusually for science, makes a literary reference when in its last line it concludes: 'Finally, it has not escaped our notice that the more we learn about the human genome, the more there is to explore.'

[7] Initial sequencing and analysis of the human genome, *Nature*, **409**, 860–921.
[8] Molecular structure of nucleic acids: a structure for deoxyribonucleic acid, *Nature*, **171**, 737–8.

6.7 The commercialisation of the genome

A scientific frisson accompanied the 2001 publication of the genome map. The Human Genome Project had started in 1990 as a publicly funded venture involving thousands of scientists sharing data and ideas. It was an important research principle of the Project that the results be available to anyone connected to the Internet.[9] However, late in the 1990s some competition emerged. A private company, Celera Genomics, was also mapping and sequencing. Using different methods, and starting much later, Celera nevertheless caught up and published its results at the same time, in the US journal *Science*. There was not much to choose between the two ventures in terms of quality of outcome, but according to the scientists of the Human Genome Project, there was a huge moral difference. When the publicly funded Human Genome Project published in *Nature*, they reminded everyone that the results could be studied simply by typing in a web address. Celera Genomics, however, put commercial controls on the use of their data. This seemed to contradict a principle of scientific research often quoted by publicly funded scientists: that experimental results should be freely available to all, so that the experiments can be replicated, challenged, and improved. Sir John Sulston called it 'privatising the genome', and compared the consequences to the British railways – whose failings were at the time being blamed on an earlier privatisation. At the Project press conference, emphasis was given to the fact that scientists in the developing world had been accessing the genome site. The implication was that such research would be stifled if payment was needed for access.

Behind the squabbles and the politics lies a deeper, and important issue. It would be false to see Celera Genomics as commercial, and to see every other genome scientist as completely without a financial interest. Across the world (but especially in the USA and the UK) small biotechnology companies are taking genome information, developing tests or drugs, and patenting them. The controversy is over what can be patented. Patents are used to protect inventions and intellectual property: once an invention is patented, you can copy it or use it but you must pay a fee (sometimes called a franchise). A patent is granted by the State to the inventor. There are good reasons for the existence of patents: apart from securing the inventor's commercial rights, a patent acts as an incentive to creative people

---

[9] The genome site is http://genome.cse.ucsc.edu/. *The Nature* genome gateway, at http://nature.com/genomics/human/, is also a useful site.

to pursue their ideas, and, if successful, to make their work available to the public.

The debate over genetics and patents arises because of the criteria used for judging patent applications. How should intellectual property be protected? The European Patent Office judges the patentability of an invention such as a new vacuum cleaner, or a new genetic test, using the same criteria. For a product to be patentable the following must apply: it should be novel; it should be inventive; it should be capable of industrial application; it should not simply be a scientific discovery; it should not be an animal or plant variety; it should not simply be a biological process (though microbiological techniques are patentable); and it must not threaten public order or morality.

There have been many successful applications for patents by companies (and universities) working with genetic material. For example a gene test may be patentable because the test does not rely only on the discovery of a gene; perhaps the test involves using lengths of DNA that occur naturally, but have been given some practical application. Similarly, a length of DNA designed for insertion in a microbe might be declared novel and patentable, because it does not exist in nature in this form. Patent applications have not been hindered by the criteria about animal or plant varieties, or 'biological processes'. A gene (or a length of DNA) is neither an animal nor a plant, and if the DNA has been substantially modified, it can count as an invention.

An interesting aspect of patent law concerns the criterion over morality. According to the European Patent Office, an invention is deemed immoral (and therefore unpatentable) if it seems likely that the invention will outrage the public. There is a good deal of room for argument here – as was seen in the case of the oncomouse, a transgenic mouse whose modifications ensure that it develops cancer. A patent was granted because the invention (the mouse) was ruled to assist in treating dangerous and widespread diseases, and poses little threat to the environment. A patent application for a transgenic animal where the modification is of mainly commercial value, such as higher yield of wool in a sheep, is less likely to be successful: in this case, the affront to public sensibilities is taken as more important than any financial advantage a company might gain.

The publicly funded research on the genome map has made it a principle that the results remain free for all to use. Yet it is important to remember that the Human Genome Project is only one part of the human genetics research programme, and will soon be complete. Other programmes, such as proteomics, will soon dominate. Perhaps the most immediate commercial pressures will be in the area of genetic diagnostics: the development of tests to identify susceptibility to a disease. No doubt many of the scientists involved in the publicly funded Human Genome Project will find themselves tempted by the commercial possibilities, or encouraged to be so by their universities.

A full treatment of genetics and patenting would be lengthy and complex. The above information is simply to show why genetic patents exist at all, and why there is controversy. Perhaps the most important point to consider is the effect that such patenting can have on the ideals of science. We have seen that science has been considered a communal enterprise, where easy access to results allows scientists to build on (or demolish) each others work. If commercial interests make it expensive or impossible to use data, then there are implications for the way that scientific research is carried out. Research tribalism will be exaggerated; public suspicion of secret laboratories or research centres will increase; and poor countries will be unable to make use of research programmes in aiding their own development. Policy makers would do well to take these possibilities seriously: the BSE affair has revealed public suspicion of experts who sound arrogant about benefits, and complacent about dangers.

6.8 Gene testing and the doctor–patient relationship

In discussing the implications of the new genetics, it is sometimes implied that one novelty is the fact that a lot is known about a bodily system, but not much about how to make a cure. Actually, this has usually been the case in medicine. Only in the late nineteenth century, when the sciences of physiology and of microbiology were complemented by safer anaesthesia and surgery, did a doctor's ability to intervene effectively grow. For centuries, anatomical knowledge had been excellent, as shown by exquisitely accurate wax models of body systems studied by eighteenth-century medical students and doctors.

Paradoxically, modern medicine is moving us towards the eighteenth-century position where the gulf between knowledge and cure was huge. For even if the Human Genome Project soon leads to a greatly increased ability to find predispositions to disease, the use of genetic techniques to adjust those predispositions remains a fantasy. In the meantime, an established moral issue will become still more controversial. For with the current limited ability to correct genetic faults, the main form of medical intervention will be abortion. Very often the person wanting a test will be someone who has experienced genetic disease in the family, and wants to know their own status or, in the case that they or their partner is pregnant, the status of their fetus; couples like this might be quite practised in discussing what they might do if their fetus turns out to have the particular impairment. Different issues will develop as the burden of dealing with unwelcome genetic news becomes more common, either if people decide to take advantage of cheap commercial tests, or if screening becomes more widespread. Screening is the type of testing applied routinely and bureaucratically to large numbers of people, independently of whether they are known to have a high genetic risk. At present, the most familiar type of screening is the ultrasound scan offered to all pregnant women. If, some day, pregnancies are screened for disease predisposition using tests involving genetic analysis, then increasing numbers of women will discover, well before birth, that their baby has a type of impairment. Such screening would not necessarily be coercive; it could become an accepted fact of reproduction that genetic screening is as natural a part of reproduction as birth itself. Yet it seems hard to imagine that the resulting information will not be burdensome. The question will be: when should a test result be taken as a justification for an abortion? At present, there has been very little debate on how this should be decided. Health professionals tend to argue that parental choice in the matter should be the most important factor, but this hardly resolves the issue. Could it be that club foot, or asthma, or late-onset heart disease, would become acceptable as reasons for abortion? Probably not, on the grounds that these conditions may not seriously affect quality of life. Society has already made some decisions on these issues. When Down syndrome is detected, abortion is offered. Yet though many Down syndrome fetuses are aborted each year, many are not, and go on to become lively and accomplished human beings. We disapprove of those who patronise people in wheelchairs, but if a fetus were discovered to have a physical disability, such as lower-limb paralysis, what should our attitude be? The official view is that the parents must decide, but there is a widely held argument that calls for caution. For if genetic tests and screening lead to an increase in terminations of impaired fetuses, what message is being sent out to the disabled? If a disabled person finds it hard to lead a fulfilled life, the reason may lie more in the prejudices of society than in the mechanics of the impairment. Parents debating the future of their impaired fetus will realise that the quality of life of the child will depend, to a greater or lesser extent, on the manner in which society views imperfection.

Another type of medical intervention will be much less disturbing. A person visiting their doctor may come to discuss their latest genetic profile: no doubt the genetic check-up will be regular, not because their genes have changed, but because new tests have become available, for new mutations. Discussions of genetic profiles, and what they signify when there is no chance of changing that profile, will lead to a redefining of the medical role. The situation at present is that most visits to the doctor are prompted by a pressing health problem that needs some attention, and the doctor tries to give immediate support. When it becomes possible for doctors and their patients to access a fairly weighty portfolio of genetic information, their relationship to each other, and to the health problem, may change. The changes involve ideas about the future, the role of doctors, and the confidentiality of their deliberations. The scale of these changes is a lively debate amongst medical commentators. What follows in the next section is a speculation of how your visit to your local doctor might change.[10]

Future doctor–patient scenarios

The emphasis of health care moves from the present to the future
Normally, people become patients when they feel themselves ill, when an ailment becomes obvious. A doctor may recognise future risk: for example someone with hypertension runs the risk of cardiac disease and stroke. The doctor will advise weight control, for example. Still, a person is only likely to be in the doctor's surgery if there is some complaint, otherwise one is regarded as guilty of 'wasting the doctor's time'. However, information from genetic tests will indicate future risk, perhaps as far as 30 years away: you will become a patient without symptoms, a patient before your time, but you will have a sense of a future that is determined, and it will be that future that you will want to discuss with your doctor.

The emphasis is on risk: medical contemplation becomes as important as medical action
When we think about our future, we conceive it in terms of probabilities. Certainly, at length we will die, but even if we have a lethal disease, we

---

[10] See Chapter 1 in *The Human Genome Project and the Future of Health Care*, edited by Thomas H. Murray, Mark A. Rothstein, and Robert F. Murray (Bloomington: Indiana University Press, 1996).

cannot be sure how long we will last or even if we will die from it: perhaps the No. 38 bus will run us over first. With genetic information we will still be dealing with probabilities, but the calculations may in time be more accurate. Furthermore, while cancers or cardiac problems might be cured, and are in some sense not the 'real you', your genetic mutations are as much you as your hair colour: the disposition to disease, and you, are one and the same.

As a result, there will be great interest in preventive medicine. Diagnosis will be easy, therapy (in the sense of complete cure) impossible. The doctor will not be a therapeutic activist, but an advisor. The skill will be in providing the right advice and information. Though cures are not available, research may provide a more and more fine-grained analysis of how environmental factors may accelerate, or delay, the onset of symptoms. This will lead to another kind of medical encounter: the contemplative interview. Through talking and listening, the doctor and patient, perhaps over many years, will track the course of the disease. There may be new drug therapies for treating the symptoms, perhaps designed specifically for particular genotypes, but the preventive measures established by the patient–doctor consultations will be important too. These consultations will be time consuming for they will seek to analyse in detail the patient's lifestyle, and for ways to delay the onset of symptoms.

Your consultation concerns not only you: a crowd of people is present in your doctor's room

When you visit the doctor, you are prepared for a private talk: a quiet consultation between an expert and yourself. It is you and your body that form the focus of attention, and of advice and treatment. We can imagine that this will change as the factual content of those conversations becomes more and more genetic. For in addition to considerations of your future, and how you might safeguard it, discussion might turn to your relatives. Your own careful deliberations might be of unusual importance to others around you, not because they are concerned for your health, but because they are concerned for their own: they have your genes too. In a real sense then, when you go to the doctor, you will take your family too.

There are others who will also be interested in your consultation, and in your doctor's advice. Insurance companies are likely to want to share in some – or much – of the genetic information that the health services are giving you. Insurance companies already gather medical information, but might declare that life and health insurance policies are not available for those with high-risk, late-onset conditions. They might also raise premiums above the average for those whom they consider to be an above-average risk. However, if they resolutely discriminate against the 'genetically disadvantaged' and refuse them insurance, they lose part of the market. Moreoever, those who have been lucky with their genetic inheritance, and know it, might themselves calculate that they can forego life, or even health, insurance. The list of those interested in your doctor's views is likely to grow. Your employers might deploy you to work in environments that would suit your genetics, and in the future, they might demand genetic information, as an extension to the general medical inquiries that personnel offices often make. Your genetics, in other words, will become part of your CV.[11]

[11] The Ethical Legal And Social Issues (ELSI) of the Human Genome Project, including the issues associated with genetic testing (codes of practice and regulations as they exist so far), and also more general biotechnology issues, are comprehensively discussed on a number of web sites, which all give useful links to other sites and publications: http://www.ornl.gov/hgmis/elsi/elsi.html; http://www.genome.gov/page.cfm?pageID=10001618; http://lawgenecentre.org/; http://www.nature.com/genomics/

7

Biology and politics

7.1 How politics enters biology

Politics is the management of society and the discussion of that management. It is about the allocation of power, and of information. Few people are distrusted as much as politicians, but even if we want to ignore politics, we must admit that society's decision makers have some impact on our lives.

Perhaps it seems obvious that disagreement and rebellion are fundamental to a proper society. Actually this is a much disputed idea. The mid-twentieth-century fascist governments of Spain, Italy and Germany did not tolerate disagreement from their citizens. When fascist governments suppress demonstrations and opposition rallies they justify themselves by speaking of the threat to society. A government that decides it has discovered the right path for society, and will not listen to anyone else, quickly removes 'the right to free association'. Opponents are not opposed by argument; they are redefined as ill, or mad, or enemies of the state. Such governments, whenever they emerge, are condemned by democratic countries for suppressing debate. Of course, it is easy to argue that many politicians in democratic countries also manipulate debate to their own ends: propaganda, 'spin' and nepotism are simply sophisticated ways of stifling head-to-head argument. These tricks by those in power are harmful because they anaesthetise politics. Politicians will not bother to argue if they know in advance that they cannot lose.

Yet in spite of the defensiveness and cynicism of modern politicians, their's is a profession highly sensitive to the world of ideas. Individual politicians may seem comically superficial, but their job is to put into practice important ideas about society. In defining what is meant by a good society, they must, at some stage, ask around. Economists, historians and military experts give them advice, so do bankers, head teachers and trade unions. Each of these groups can be relied upon to put their own view: but what about scientists? Their job is to investigate nature, not come up with a 'view'. Can science ever advise on what a better society should be, and how it could be achieved? To behave like that is to enter the world of politics. This chapter looks at a few examples of biology mixing with politics. We shall see some of the dangers. I hope that the arguments we examine will make you cautious about scientific advice given to governments, but will nevertheless encourage you in thinking that political debate is important.

To begin, we should briefly consider one reason why biological ideas might be used in politics. Humans are animals, whose evolution depended on Darwin's mechanism of natural selection. Our physique, metabolism and intelligence could not exist without the action of genes. That much is not controversial. It is the debate over the way genes do, and do not, determine behaviour that is interesting politically: whether these aspects of human life are heritable – that is, whether some of the variation between humans is owing to genes. If politics is considered to mean the debate over how power is distributed within a society, then it is clear that politicians will from time to time take seriously ideas about inheritance, and its role in defining the different groups within society. I shall describe in this chapter examples of how biology has been used to justify discrimination, against blacks, or Jews. We might protest that the kind of biology used to justify racism is 'bad biology'. It is good, therefore, to revisit some aspects of twentieth-century history, such as the Nazism of the 1930s, and reflect on how Hitler's ideas on genetics do not seem to have been particularly different from mainstream genetics in the rest of Europe or America.

Today only the most innocuous work on genetic differences between ethnic groupings is allowable. Contemporary human geneticists, understandably, stress their role in understanding disease, not in understanding society. Towards the end of this chapter I will explore briefly a subject also mentioned in Chapter 3: evolutionary pyschology. This field, once better known as 'sociobiology', generates fierce debate because it aims to explain human behaviour by reference to Darwin's theory of natural selection. By exploring the links between Darwinism and human behaviour, evolutionary psychology necessarily draws on ideas about genes. Moreover, it is a field that seems drawn to sensitive areas: homicide, stepchildren, sexism, aggression, love. As we shall see, however cautious an individual evolutionary psychologist might be, the field's ambitions in relating genetics to complex social problems awaken fierce controversy.

An example is the investigations of evolutionary psychology into sexual relations between humans. Sex education in school biology lessons emphasises the mechanical aspects of sexual reproduction, but leaves the cultural and emotional aspects to lessons such as social education, or religious education. According to evolutionary psychologists, biology need not feel bashful in providing clues about relations between men and women: can it, for example, explain sexism? Sexism in the workplace, whether by individuals or by the institution, is frowned upon by enlightened governments, and legislated against. Might biology have a view? We can start by looking at the biology of mammalian reproduction. We notice that viviparity correlates with a female investing more in her young than does her mate. A male can best increase his genetic representation in the next generation by impregnating a number of mates: to use human terms, adulterous behaviour in males will be encouraged by natural selection. Might it, therefore, be reasonable to excuse promiscuous and faithless men by pointing to the teachings of biology? This would depend on our acceptance of two things. Firstly, that we accept as reliable the admittedly orthodox view that males in the animal kingdom are more promiscuous than females; and secondly, that it means something to say that amongst the turmoil of sexual relations between men and women, the 'genetic message' is clear, unequivocal, and influential. I will give a brief review of this particular example at the end of the chapter.

In evaluating all such arguments, an enormous number of issues must be considered. At this point I will consider only two. The first is called the fallacy of reification: the mistake where something complex and vaguely drawn is treated as a simple, concrete entity. The problem is simply this: we might worry about sexism, but can we actually define it as a single thing? A simple definition is going to be needed if its roots are going to be traced to our mammalian evolution. Yet sexism may not be a simple behaviour, but a vast network of influences that weave through institutions and people. We might agree that it is bad; but we might not agree on what 'it' is. Trying to force a definition may be no good at all, and simply give a false sense that a problem has been made amenable to the examinations of science. The second problem is that of self-delusion, and concerns the security of the science that is used to explain the human activity. We have seen in this book that biologists argue about their data. Interpretations of data change over time, and the information itself may change. A relevant example can be drawn from primate studies. During the 1960s and 1970s, studies of primate behaviour began to show that female chimps, contrary to earlier opinion, were active sexual partners: they did not simply hang around waiting for something to happen. No longer were the males the dominant sexual actors; females were involved too. How could this be? It is not likely be that ape behaviour had changed; it is much more likely that the scientists had adjusted the way they collected evidence. This development, in turn, depended on the fact that women primatologists had become more common, and also that the women's liberation movement was making its impact on society. The assumption that female chimps did not do much except bring up baby was not one held by the new women scientists.

Reification and self-delusion are the two important steps in using biology (not necessarily evolutionary biology) to make a political point. Firstly, you make the links between biology and human behaviour superficially plausible by reducing behaviour to a cluster of over-simple, definable concepts. Secondly, you are highly selective in your evidence, interpreting the biological evidence as sure support for your own particular account of the causes of human behaviour. Sexism, therefore, is sometimes portrayed not so much as a social pathology, but as a leftover from our evolutionary past: the emphasis is placed on the latter, rather than the former. Just as, allegedly, in our animal past it was the females who bore the young and took charge of parental care, so today it is best (more natural) if women think mostly of their children, their appearance, and their home. In this 'biological interpretation' the emphasis runs as follows: however much we disapprove of sexism, it exists as part of human nature, thanks to our evolution. Our legislation and fine words will always be in opposition to our real selves. Political reform and equal opportunity regulations, working against the grain of inbuilt human nature, will, therefore, prove unsuccessful and futile. I have deliberately chosen a simple example to illustrate how you can mount an attack on this type of argument, by suggesting that sexism in the modern world cannot simply be compared with investment strategies in the animal kingdom, and that when it comes to the evolutionary past of humans, we have very little evidence of how our ancestors behaved. Fossils are not good at revealing behaviour, but they are excellent at receiving the projections of contemporary ideology.[1]

[1] Sexism is discrimination between people on the basis of their gender. Clearly then, the fact of sexism, and the fact that many women bring up children and work, are separable. The sexism would arise if it was argued that women ought not both work and have children, and should be discouraged

7.2 The politics of Darwin

When Charles Darwin reached the final pages of his book *The Origin of Species* he wrote: 'much light will be thrown on the origin of man and his history.' In his usual laid-back way, Darwin was once again saying something disconcerting. No one could escape the force of Darwin's primary point – all life, even human life, had arisen from a simple life form. Darwin was clear in his own mind that natural selection was as much a cause of the evolution of people as of other organisms, but he was deliberately cautious. His great battle was to persuade others of the fact of evolution. The conclusion that humans too must have evolved, and would show in their bodies and minds the effects of natural selection, could be left to others to draw. In time, Darwin did return to the theme. As old age approached, and with many other biologists speculating on the implications of Darwinism for human society, Darwin published in 1871 *The Descent of Man*.[2] Most of that book concerned the role of sexual selection; but he already saw that his views on animal evolution, if applied to people, had political implications. He wrote 'With savages, the weak in body or mind are soon eliminated; and those that survive commonly exhibit a vigorous state of health. We civilised men, on the other hand, do our utmost to check the process of elimination; we build asylums for the imbeciles, the maimed and the sick . . .', and concluded 'thus the weak members of society propagate their kind'.

The debate about the difference between people and other animals can be approached in many different ways. There may, for example, be a religious implication if the differences between humans and other animals are questioned. Ethical issues arise too: if the divide between animals and humans is artificial, how shall we best justify vivisection? Scientifically we can approach the question by exploring similarities in skeleton, in physiology, and in genome. Darwin's ideas suggest a slightly different tool: that of evolution. After Darwin, it became reasonable to ask whether natural selection alone could produce the moral sensibilities of people, or their musicality. Similarly, if society is simply a large group of people, each of whom is only a kind of animal, then maybe (using Darwinian ideas) biology can explain the human events we see around us, and even advise politicians on what should be done. Ever since Darwin, as part of this

---

from doing so by poor pay and conditions. The heat around evolutionary psychology arises because of the field's inevitable association with the following argument: females invest more in their offspring; in women, this translates as taking charge of childcare; therefore, biology teaches us that women are best off if they stay at home and do not work.

[2] *The Descent of Man, and Selection in Relation to Sex* (London: Murray, 1871; 2nd revised edn, 1874).

debate, there have been many attempts to find – or detonate – a dividing line between humans and the other inhabitants of the planet. Language, art, the kind of intelligence that leads to philosophy and biology: each can be put forward as candidate criteria. The existence of freewill, that strong feeling each of us has of being a 'person' capable of making decisions and acting as we wish, is also used to separate us from animals, or at least to suggest that we are not utterly enslaved by genetics, or the environment. The issue is as alive as ever today, as is obvious from our interest in the related question that concerns the difference between humans and machines. The exploration of these questions is not our task here, but we should note the fundamental point: that Darwin undermined the notion of a clear dividing line, and ushered in an age where priests and politicians regularly looked to the animal and plant kingdoms for ideas about human society. Thus it was that Darwin's theory of natural selection leaked into nineteenth-century political debate. At a time when the theory was still viewed with some scepticism by biologists, its message about 'the survival of the fittest' was being taken up with enthusiasm by social commentators, who saw in Darwin's law an awful warning: to protect the weak is to override natural selection, the process whose finest achievement is the human race.

Darwin himself had pointed out that animal breeders would soon go out of business if they were as irrational in their choice of cross as the average human. He thought that the human equivalents of prize pedigree bulldogs and labradors were sluggish at reproducing. Conversely, the mongrelly, the rough and the pox-ridden, kept alive by charity and health measures, were breeding far more quickly. Humans had reached the pinnacle of evolution, thanks to natural selection, but were now turning their back on a basic guiding principle of nature: that the weak must perish. Instead, the weak were being kept alive. Though nineteenth-century cities were a symbol of progress, and were marked by great monuments to mechanical and imperial power, they were also remarkable for their crime, pollution and ill health. To the middle class, working hard (in their view), and prospering, it was an appalling sight. To them, Victorian society was at risk of degenerating. It seemed possible that humankind had literally put evolution into reverse. What was the use of technological progress if it was accompanied by – perhaps even caused – an increase in ugliness and despair? Moreover, this was a time when everyone, professional biologists included, considered it plainly the case that poverty and unemployment and crime are heritable characteristics, passed down the generations. Not that there was a theory of genetics. Around 1870, not even Darwin knew about genes, but everyone could see that bad blood, manifesting itself in scaring crowds of badly behaved and illiterate children, was spreading fast by the uncontrolled breeding of undesirables. The most feared antisocial trait was 'feeble-mindedness'. Again, it was certainly inherited, and for those afflicted, the problems of poverty, homelessness and unemployment were almost inevitable. Anyone could see why feeble-mindedness was on the increase. Feeble-minded women, of course, cannot 'stop themselves'. They have child after child, each by a different man. Each inherited the basic problem of blood; each was first neglected, and then ruined, as the feckless mother spiralled further into degeneracy.

7.3 Eugenics: the genetic improvement of people

Fears about the quality of the nation's breeding stock prompted the birth of a new science – 'eugenics'. The term comes from the Greek *eugenes*, meaning 'good birth'. Eugenics was founded by Francis Galton, an academic working from University College London. His real love was counting, and it is to him that we owe the establishment of statistics. However, like many Victorian gentlemen, he was disturbed by the degenerate people living around the railway arches and the markets. His love of counting prompted him to apply science to the widespread concern about degeneracy. He would measure the problem, and suggest a solution; but in those days there was neither statistics nor genetics, so Galton had to start from new. In one of his investigations, he studied eminent British families, and noticed a disproportionate number of talented individuals. When one of these important families sat down to eat their turkey on Christmas day, very likely there were top scientists, cabinet ministers and army generals passing round the gravy boat. Clearly, brilliance, perseverance and good manners ran as truly in the blood as did prostitution and alcoholism, but, like Galton, these brilliant fellows did not leave enough offspring – elites do not reproduce well. For Galton, then, eugenics would be a movement that would use science to change this situation.

How could this be done? Little was known about the workings of inheritance. Galton experimented with plants, but found discouraging results. When he bred from the tallest plants within a population, the offspring tended to revert to the normal. This suggested that two eminent parents would tend to produce fairly average children. Galton did not have a clear idea about how these problems could be overcome; but he was adamant that the rich and talented should be encouraged to do their bit for the next generation. This was 'positive eugenics', the brand that aimed to accentuate the good. However, for those who followed Dalton, negative eugenics (the destruction of the bad) seemed more practical. The rich do not take kindly to being told by the state what to do, especially when it comes to their sex life. The poor, the sick and the feeble-minded can be pushed around much more easily. Eugenics grew into a powerful doctrine, but one whose concerns were with society's less favoured citizens. We shall see that these worries eventually came to be used to justify segregation, racism, sterilisation and, finally, murder.

In the nineteenth century, the biology used by eugenicists at first favoured a rather generous attitude to those who were feeble-minded, or idle, or inclined to steal. For until about 1880, those who wanted to view the degenerate classes favourably, and put in place reformist measures that might improve housing and education, found biology on their side. The view of inheritance then widely favoured was that of Lamarck. Broadly understood as implying that changes in one generation could be inherited by the next, there seemed in Lamarckism the prospect of improving bad blood over several generations simply by a programme of education. If parents stopped their reckless behaviour, and instead cultivated generous feelings, they would actually pass to their children an improved 'seed'. Such children would be born with a wholesome temperament and perhaps make an important contribution to society. In this belief lay some idea that improved parents actually passed on to their children the better characteristics. According to this biology, then, free education for all and better housing were worthwhile. People could be transformed, and with them their descendants. Meanwhile, with science now a professional business, with university departments and professors, the inheritance of dysfunctional traits could be investigated. In the late nineteenth-century 'Jukes Study', we see how the same scientific data can be used to support contradictory political stances.

"The Jukes": a Study in Crime, Pauperism, Disease and Heredity was published in 1877 and gives details of a family of famous degenerates. The author, Richard L. Dugdale, was on the board of the Prison Association of New York, and he interviewed prisoners in 13 jails. In the process he came

across a family, given the pseudonym 'Jukes', whose members managed a collective criminal career that spanned several generations. The men of the Jukes family were murderers, rapists, burglars, wife beaters and alcoholics. In each generation the same worrying traits emerged. Here again was evidence of the heritability of antisocial behaviour.

However, for Dugdale, writing before the birth of modern genetics, the lesson of the Jukes family was not a cause for despair. Following Lamarckian ideas, he saw in the Jukes not the hand of fate, but the chance to show the effectiveness of liberal politics. He believed that education and training, consistently applied to young people, will not only affect their cerebral tissue, but also their 'heritable matter'. With a biology like this, the political implications are clear: 'public health and infant education . . . are the two legs upon which the general morality of the future must travel'.

Biology was changing. The partnership between politics and biology was to shift towards a much more punitive, and even vicious, stance. In 1883, the German cytologist August Weismann had distinguished between two kinds of cells making up the human body: germ cells, in the testes and ovaries, and somatic cells (the rest). He argued that the germ cells are completely isolated. It is their genetic material that is passed down from one generation to the next; this cannot be influenced by the somatic cells. Thus, whatever happens to the muscles (in the case of the blacksmiths) or the cerebral tissue (in the case of criminals converted to a life of moderation), the germ line remains unaltered. This shift in biological reasoning led to a eugenics interested in discrimination, rather than reform. From now on the human targets of eugenical ideas would be under greater threat.

British, American and German scientists became convinced by Weismann's doctrine. It was still taken for granted that feeble-mindedness, licentiousness and idleness were heritable; but now, thanks to Weismann's doctrine, these defects looked like indelible stains. In terms of politics, the implications were thought provoking. Reform of such people was useless: they were fated to be as they were. The important thing would be to protect the future of society by preventing them from breeding. Indeed, charity and education would simply be a waste of money, for though the individual might be helped in some small way, the inheritance remains unreachable, and the problem will soon manifest itself as strongly in the next generation. Perhaps those afflicted by genetic conditions such as 'idleness' should simply be segregated away in asylums, locked in single-sex wards, and kept celibate.[3]

7.4 Measuring intelligence

At the start of twentieth century, asylums for the feeble-minded were common, and full. Policy makers became aware that they were expensive to run. If the aim was simply to prevent people breeding, then sterilisation would be as effective, and much cheaper; but how many feeble-minded people could be found, for example, in the USA? If the prevalence of feeble-mindedness in society was to be monitored, and, better, reversed, it would have to be measured. The skill, of course, is in devising a reliable test. In this sense, while measuring height is easy, measuring mental ability is more difficult. What we nowadays call 'intelligence' resists easy definition. Intuitively, we know what we mean by intelligence, and feel that we recognise it when we see it; but to devise tests that can be applied to any human being, especially if the tests will decide their future, is much more controversial. In the middle of the nineteenth century the science of phrenology, or 'measuring heads', had been popular. Cranium size, and even cranium shape, were both held to be important indicators of intelligence and personality. At a time when the different peoples of the world were often held to belong to different species, brain measurement might even show sharp, definable differences between such 'species'. In one case, Samuel Morton, a scientist and physician from Philadelphia, began to collect human skulls; by the time of his death his cupboards were filled with over a thousand. Morton tested each one by filling it with mustard seed, tipping the seeds back into a measuring cylinder, and reading off the result in cubic inches (later he used lead shoot as it gave more consistent results). The original owner of the skull was also noted: English, Native American, Negro. According to Morton, his 'objective results' showed some clear differences: whites had the largest brains, blacks the smallest. Morton's results are now considered completely unreliable, revealing

---

[3] With so many possibilities, frightening or otherwise, stemming from the Human Genome Project, scholarship into the history of eugenics has suddenly come to seem of urgent importance. For an excellent, accessible account of eugenics and its modern significance see Diane B. Paul's *Controlling Human Heredity: 1865 to the Present* (Atlantic Highlands, N.J.: Humanities Press International/ Prometheus Books, 1995). The problems and fallacies of measuring human intelligence are exposed to fierce scrutiny in Stephen Jay Gould's *The Mismeasure of Man* (New York: W.W. Norton 1981) – my accounts of intelligence measuring and testing in Section 7.4 are taken from this source. A revised and expanded edition of Gould's book was published by W.W. Norton in 1996.

much more about his own preconceptions than anything real about intelligence, ethnic groups, or even brain size. In a study of how Morton's results might have been biased by his expectations, the writer and scientist Stephen Jay Gould imagines Morton at work. 'Morton, measuring by seed, picks up a threateningly large black skull, fills it lightly and gives it a few desultory shakes. Next, he takes a distressingly small Caucasian skull, shakes hard, and pushes mightily at the foramen magnum with his thumb.' Gould is suggesting here that, even unconsciously, Morton packs the Caucasian skull harder than he does the black skull, and so achieves a higher volume.

The attempts in the twentieth century to measure feeble-mindedness (or intelligence) also struggled with its methods, and earned many critics. The two problems with Morton's work were his identification of human brain size as an important marker, and his actual methodology. Even if his method of measuring brain volume had been reliable, his initial assumption that brain size in humans indicated something important, in the end was soon seen as false. Could the twentieth-century eugenicists manage something better? In France, the psychologist Alfred Binet was beginning to look for reliable ways of measuring intelligence. He had already found that measuring skulls produced differences far too small to be of use, and had also spotted a problem with making the measurements. 'I feared that in making measurements on heads with the intention of finding a difference in volume between an intelligent and a less intelligent head, I would be led to increase, unconsciously and in good faith, the cephalic volume of intelligent heads, and to decrease that of unintelligent heads.' So when in 1904 Binet was commissioned by the French Government to find ways of identifying slow learners in the school population, he chose a test made up of some basic tasks in reasoning, comprehension and correction. They were quite diverse, and Binet hoped that by having a broad range of puzzles, he would be able to deduce from the results a fair idea of a child's ability. The results became the 'intelligence quotient'. The tasks ranged in difficulty, and each was assigned an age level – the age at which a normally intelligent child would be able to do the task. A child doing the test eventually stumbled at a question, or series of questions, marking a particular age level. This was viewed as his actual mental age. By dividing this age by his real age, and multiplying by 100, the intelligence quotient (IQ) is calculated. An IQ of 100, therefore, is average.

Binet was clear that his tests were to be simply used for identifying some problems in part of the school population. He did not see them as a general 'intelligence' test. He warned that intelligence is not like height, and he worried that his test could be abused by school masters: 'They seem to reason in the following way: "Here is an excellent opportunity for getting rid of all the children who trouble us," and without the true critical spirit, they designate all who are unruly, or disinterested in the school'. This quote shows that the originator of the intelligence test was highly enlightened, and measured, about the value of this work. He aimed to identify some children so as to help them, not so as to condemn them, but he understood too that his invention was likely to be abused. He claimed that teachers 'are not interested in students who lack intelligence. They have neither sympathy nor respect for them, and their intemperate language leads them to say such things in their presence as "This is a child who will never amount to anything . . . he is poorly endowed . . . he is not intelligent at all"'. For those in America convinced that feeble-mindedness was a major social problem, the intelligence test was, however, a fantasy that could not be resisted. The test was adapted and enlarged, and declared to be a fair judge of intelligence. Moreover, the test could be given, and the results processed, quickly enough for large-scale testing to be a practicality.

Just after World War I, 1.75 million soldiers in the US Army were tested. The results of the US Army tests showed that the average mental age for white soldiers was 13.08 years; for blacks it was lower still. Not only was this a frightening indication about the army: if the American population in general had an average mental age of 13, what hope could there be from this celebrated democracy? Worse still, clear differences had emerged once again in the intelligence of different races. Could not the results demonstrate something about environment? No, because the tests, it was assumed, tested native, i.e. inborn intelligence. These army tests were the first written IQ tests to be accepted as reliable. They indicated something important about a person's inherited intelligence, and they were easy to set, and to mark. Of course we now know that they were culturally biased, and discriminated against those unfamiliar with middle-class American culture.

7.5 Eugenics and genetics

Eugenics was not without its critics. Right from its inception in the late nineteenth century, there were those who saw the movement as coercive. In the UK especially, trade unions became suspicious. They saw eugenic ideas as just one more example of the rich man's desire to bully the poor. The Roman Catholic Church was doctrinally opposed, on the grounds that eugenics interfered with reproduction, and, therefore, with life. From our modern perspective we might ask: how did science, especially developments in genetics, impact on eugenics? Consider, for example, the rediscovery of Mendel's work in 1900. His work on peas showed how characteristics can be controlled by 'factors' that operate as pairs. The terms 'dominant' and 'recessive' were introduced, and it became clear why an inherited characteristic persists, yet does not necessarily show in every generation. Mendelian genetics would tend to suggest that, supposing for the moment that feeble-mindedness is recessive, most feeble-minded people would be born to normal adults. Moreover, if feeble-mindedness is a recessive trait, then sterilising feeble-minded people will miss most carriers of the so-called gene. Some commentators though chose to see in Mendelism only the simple point that genetic characteristics endure. In 1927, W.R. Inge, Dean of St Paul's, put it like this in his aptly titled *Outspoken Essays*:

> Feeble-mindedness follows simple Mendelian rules. It cannot be bred out of a family in which it has established itself, but it could be eliminated by bringing the infected stock to an end. Unfortunately, the birth of the feeble-minded is quite fifty percent higher than that of normal persons. Feeble-minded women, being unable to protect themselves, often have an illegitimate child nearly every year.[4]

Inge, however, recoiled from advocating sterilisation, and infuriated the eugenicists by describing it as 'mutilation'. Yet even amongst leading geneticists, there was a difference of opinion between supporters of eugenics, and those who doubted its scientific credentials and its motives. The socialist biologist J.B.S. Haldane was of the opinion that the growing science of heredity was being used in Britain to support the political opinions of the extreme right, and in America by some of the most ferocious enemies of human liberty. For such people, the whole of eugenics was based on an anti-poor bias, but given a scientific veneer. Attacks on eugenics questioned the existence of the trait of 'feeble-mindedness', and undermined the credentials of ventures like the Army tests by pointing to cultural bias. Even if mental ability has a genetic component, the method of inheritance would not be along simple Mendelian lines, but

---

[4] *Outspoken Essays (2nd Series)* (London: Longmans, Green and Co. Ltd., 1927).

would be polygenic, controlled by many genes. Opponents of eugenics pointed out that almost nothing was known about such inheritance, and no sensible social policy could be derived from the science. Yet, even with feeble-mindedness becoming a controversial concept, there were many distinguished geneticists willing to support the eugenic movement. R. A. Fisher, who updated Darwinism to take account of Mendel, calculated in the 1930s that the sterilisation of all the feeble-minded of one generation would lower its incidence, in one generation, by 36%, and pointed out that this was a result that no one who cared for the future of any country could afford to ignore.

Eugenics portrayed itself as science, and in this way gained authority. It is important to realise how, at the start of the twentieth century, the achievements of science were becoming truly impressive. Since the late nineteenth century, inventions included X-rays, fertilisers, numerous new drugs, cinema, radio and phonographs. So, at the start of the twentieth century, the powers of science were clearly evident, and if science and technology could improve agriculture, very likely the science of genetics could similarly improve the human race. It would be perverse to state that life was perfect. Surely in the improvement of life, scientific breeding could very likely play its role. For biologists, this might help them take their place of importance in a society already transformed by the applications of chemistry and physics. The lure must have been quite strong. Eugenics was the area where geneticists could give their advice on how life should be lived. In his book *In the Name of Eugenics*,[5] the historian Daniel Kevles quotes one excited biologist: 'The world is to be operated on scientific principles. The conduct of life and society are to be based, as they should be, on sound biological maxims! ... Biology has become popular!' Writers too saw hope in what they called the 'scientific society'. H.G. Wells and Aldous Huxley wanted breeding to be controlled. For them, this was socialism in action: if the preparation of a better, more equitable society involved a coercive act by the state, then so be it. The state is entitled to act on behalf its people. Huxley wrote: 'we know enough, thanks to Mr Fisher's admirable work, to foresee the rapid deterioration, unless we take remedial measures, of the whole West European stock'.[6]

[5] *In the Name of Eugenics: Genetics and the Uses of Human Heredity* (Cambridge, Mass.: Harvard University Press, 1995).
[6] 'Science and civilisation' in *The Hidden Huxley: Contempt and Compassion for the Masses*, edited by David Bradshaw (London: Faber and Faber, 1994), p 112.

From the beginning of the twentieth century, individual countries began to legislate for sterilisation. In the USA, the laws were enacted by individual states. In Europe, countries passing sterilisation laws included Germany (1933), Norway (1934), Sweden (1934) and Iceland (1938). The dates are suggestive: these were the years of economic depression, when the expenses of an asylum were most likely to be resented. Though science might be interpreted as implying that sterilisation might have little affect overall on the gene pool, the idea of feeble-minded people becoming parents was too frightful to contemplate. Asylums were too expensive; sterilisation was by far the cheaper option. The ethical problems of compulsory surgery seemed trivially unimportant when compared with the menace of the spread of feeble-mindedness.

In America, another factor became decisive. Years of immigration had led to a fairly heterogeneous society. It became fashionable to enquire whether America's genetic excellence was at risk of being diluted by immigration: in particular, were there some individuals, or groups of individuals, who should simply be barred from entering America? Not surprisingly, racism and eugenics linked up to implement aggressively racist immigration laws. Eugenics had been defined from its beginnings as about 'improving the genetic stock'. Inevitably it was the nation's genes that might be in danger; equally inevitably, the protection of that heritage would lead to xenophobia and discrimination. Yet there was a complicating factor. Inbreeding was known to be harmful; hybrid vigour was a common phrase of the plant breeder. Why might this not apply to 'miscegenation' (the breeding together of the 'races')? The question was: which genes ought to be thrown into the melting pot? The Army tests themselves suggested that 45% of foreign-born draftees had a mental age of less than 8 years (the native-born figure was 21%). The Harvard geneticist Edward East wrote:[7] 'The Negro race as a whole is possessed of undesirable transmissable qualities both physical and mental, which seem to justify not only a line but a wide gulf to be fixed permanently between it and the white race.' Eugenicists in American worried, therefore, that the immigrants arriving in New York might be bringing the wrong kind of genes with them. When it came to contributing to the American gene pool, or even being in the country, blacks were considered less suitable than whites. As Edward East put it 'The ingredients in the Melting Pot

[7] Diane B. Paul, *Controlling Human Heredity: 1865 to the Present* (Atlantic Highlands, N.J.: Humanities Press International/Prometheus Books, 1995), p. 111.

must be sound at the beginning, for one does not improve the amalgam by putting in dross.'⁷ As with racists today, those already inclined to distrust immigrants will not be enlightened by the findings of science. There were scientists who challenged colleagues' views that claimed interracial breeding was harmful; but while scientists disagreed amongst themselves about 'mixed breeding', American politicians accepted that the American race could suffer from too generous a mixing with other races. There was a phrase for this: race suicide. Eugenics, however, is not simply theory: it is also action. For eugenics to impact on misceganation, there had to be immigration control. The 1924 Restriction Act, limiting immigration, was justified as a eugenical plan: for any immigrant group, the number allowed into the USA each year was restricted to a figure equivalent to 2% of the number resident in America in 1890 – a time when immigration was low.

7.6 Biology and the Nazis

The Nazi party was interested in eugenics well before Hitler became Chancellor of Germany in 1933. Throughout most of the 1920s the Nazis had been a relatively insignificant force in German politics, but when the economy collapsed in 1930, Hitler's authoritarian message began to win votes. He demonised Jews as 'anti-German', and complained that 'world Jewry' was conspiring against the national interest. Many other groups disturbed him as well: in time the Nazis would try to rid Germany of the mentally and physically disabled, depressives, alcoholics, homosexuals, gypsies and communists. Hitler had read the eugenics literature, especially that from America, and was anxious to use eugenic measures to sterilise those with inherited disorders.

It is worth considering how science in America and Nazi Germany interacted, for in 1933 Hitler's inclinations were quite clear. It was in that year when he took power, that his government passed the 'Law for the Prevention of Genetically Diseased Progeny'. Sterilisation was made compulsory for anyone, inside or outside an institution, who suffered from 'congenital feeble-mindedness, schizophrenia, manic depression, severe physical deformity, hereditary epilepsy, Huntington's chorea, hereditary blindness and deafness and severe alcoholism'. It is estimated that between 320 000 and 400 000 Germans were sterilised, mostly before 1937. During the years 1933–37, Hitler's policy was widely reported in America, and widely admired. It was the thoroughness of German policy that was so striking. The laws in America simply legalised sterilisation. The Nazi laws were more systematic. They set up an extremely thorough system that aimed to identify and sterilise anyone with a congenital defect. Doctors were required to register cases of genetic disease, so that the authorities could then proceed to sterilisation. The programme of sterilisation was accompanied by a propaganda campaign: many of the people being sterilised lived outside institutions with their relatives, and families needed to be reassured. The propaganda campaign drew heavily on the many scientific articles in American journals expressing admiration of the German programme.

The sterilisation law passed in 1933 was directed against the disabled. There were also various measures passed to promote the 'right' kind of marriages. The Nazi goal of racial hygiene was necessarily racist, and throughout the 1930s discriminatory policies were implemented against German Jews. Once again, to prevent European and American criticism of German policy, the Nazis put great efforts into national and international propaganda justifying anti-Semitism and racism. The Nazis honoured foreign scientists with honorary degrees, and were pleased to host international eugenics symposia. In this way, they won some support for their policies, and could declare to the German people that there was international acclaim for Nazi biology. On their part, American eugenicists played down the racist aspects of the Nazis, and encouraged the idea that Nazi ideology was science in action. For example a president of the American Eugenics Society, Frederick Osborn, criticised Nazi discrimination against Jews, but described the eugenic programme as 'the most important experiment which has ever been tried'. He described the sterilisation programme as 'apparently an excellent one'.[8] Only in 1938, when the belligerence of Hitler was most obvious, did the intellectual support for Germany finally fall away. By that time, with the expenses of war a priority, the Nazi sterilisation policies were mutating: before long, the asylums were being emptied, and their inmates sent to killing centres; homosexuals and gypsies were simply being shot. By 1943 eugenic practices had

[8] For a full account of the international dimensions to pre-Second World War eugenics, see Stefan Kühl's *The Nazi Connection: Eugenics, American Racism and German National Socialism* (Oxford: Oxford University Press, 1994). Recently, scholars have looked into the fate of Nazi doctors and scientists after the Second World War, and have shown that in the main they were quickly back at work in German universities, in spite of world revulsion to the Holocaust. It has been pointed out that this rapid rehabilitation was partly owing to Western fear of another political problem, namely the Soviet dictator Josef Stalin. Like Hitler, Stalin was also interested in biology. His purge of Darwinian geneticists, in favour of a state approval of the Lamarckian scientist Lysenko, doubtless suggested to the West that German genetics needed to be back at full strength – whatever its unpleasant past.

transformed into the century's most brutal example of mass killing, the murder of six million Jews in the death camps of Auschwitz, Buchenwald and Belsen.

Nazi atrocities became widely known only when allied troops entered the death camps. The trauma of the camp survivors, and the lives of their companions who died, are today marked across the world by holocaust memorials, by exhibitions, by research institutes, by art and literature, and by attempts to arrest and bring to trial those suspected of war crimes. The inquest began soon after the war ended, with the Nuremberg trials. In the dock were a range of Nazi leaders, army officers and camp commanders (though not Hitler, who committed suicide in his bunker in Berlin). During the trial interviews it became clear that Nazi atrocities had been organised in a systematic, bureaucratic, scientific manner. The death camps were the end – and the goal – of a brutally efficient system: one that involved the rail network, chemical manufacturers and incinerator designers, as well as the military and the full bureaucratic operations of the Nazi government. The Nazi sense of discipline and order, admired by many in Europe and America before the war, had in the end been instrumental to the catastrophe. Eugenics, in the 1930s, was one of Nazism's 'scientific philosophies', attracting attention and praise. Now, with the emerging details of the murder of asylum inmates, homosexuals, Jews, gypsies and political opponents, eugenics – so innocuous in its definition as the 'planned improvement of the human race' – looked very dangerous indeed. The word stopped being used: eugenics was an embarrassing anomaly in the history of science, a disastrous blurring of the boundary between science and politics. There were no more eugenics conferences, and for 30 years, human genetics became an obscure branch of medicine, its attention focused on a few illnesses.

Today the ideas of eugenics are once more being debated. There are many reasons for this. Firstly, the Human Genome Project, and its associated promise of genetic adjustment, raises questions about how this knowledge should be used, and, in particular, who should make the decisions. Secondly, historians have, since the 1970s, revealed how smoothly eugenics was integrated into mainstream science, and into wider culture; it was not simply a campaign run by a set of misguided enthusiasts. Thirdly, over the last 30 years, a series of developments have made reproductive control seem normal: the contraceptive pill; fetal testing (amniocentesis or chorionic villus sampling are common interventions in a pregnancy); and legalised abortion.

From its inception, eugenics has been associated with coercion. When life scientists today agree that modern genetics has eugenic implications, everyone is careful to emphasise that the decisions will be made by individuals, not by the state. Thus if we are entering a new world of genetic modification of humans, we will be looking at eugenics by choice, not by force. Why should there be any problem with eugenics, once we have reassured ourselves that choices rest with the individuals, not the state? The developing knowledge of human genetics will come to include an array of tests for abnormalities in the fetus, the embryo, and even the sperm and the egg. There will be an increased opportunity to prevent the birth of babies who have, or will develop, diseases and impairments of genetic origin. Even quite complex problems that have a polygenic inheritance, such as some heart disease, may come to be the subject of genetic testing. Though germ line therapy is currently banned, the eventual possibility that genes could be altered, removed or replaced, with the changes running down through the generations, is radically eugenic.

No one would dispute how severely lives can be affected by genetic impairment. Equally, it is clear that 'disability' is a broad term. For example people's attitude to Down syndrome varies. Some prospective parents are prepared to raise a child with the disability; others are not, and the choice remains with the parents. As the number of choices available to parents multiplies, so will the variety of attitudes to disability. It is possible to imagine a society where disabled people are fully accepted, and another society that might instead discriminate against the disabled, and frown upon those who in spite of all the available medical interventions, bring into being a disabled child. It might seem obvious that disability is best prevented, but when a society has the tools to reduce it, how will it view those who, despite all scientific developments, are born disabled? Clearly, the attitude of society will affect the choices made by parents. If a pregnant women finds that her baby will suffer from a mild disability, her choice regarding abortion could be affected by her perceptions of available support. In this way, coercion returns to genetics.

Personal choice is never completely free. Conventions about bringing up children influence the daily behaviour of families, and their development. Conventions about family life are set to extend back in time, to fetal development, choice of embryos and health assessment of eggs and sperm. These conventions may come to influence decisions about embryos apparently likely to become individuals with maladies not necessarily associated with disability at present: asthma, eczema, diabetes and heart disease. The implication is that eugenics will return as a public controversy, and will require political, as well as personal, decisions.

7.7 The world view from evolutionary psychology

If eugenics is an attempt to overcome the problems we inherit, then evolutionary psychology is an attempt to come to terms with them. This time it is not inherited single-gene or polygenic disorders that are the object of debate. It is human emotions and behaviour: our psychology. For evolutionary psychology takes the view that our psyche has evolved by Darwinian processes: that the way we behave from day to day can best be understood as having an evolutionary history. Our psyche received an indelible evolutionary stamp long ago, when early humans were hunter–gatherers living in stringent conditions of natural selection. At the same time as the brain expanded, the mind too was enlarging its scope. The complexities of living in groups, of bringing up families and of communicating by talking were being encompassed by these early humans. Natural selection would ensure that successful psychological attributes – adaptive to early human environments (which included other humans) – would persist and spread, and in large part continue through to modern life. According to the world view of evolutionary psychology, we might expect then that aggression, for example, or the ability to lie, or sexism, are each explained not solely by reference to the modern social context, but by Darwinian evolution. If this is true, then there will be implications for those who deal with the practicalities of human behaviour, such as politicians, policy makers, and even town planners.

The evolutionary psychologists are doing more than merely asserting that evolution has had a hand in developing our mind, though of course that is true: the brain throughout its history has been subject to biological constraint. Evolutionary psychologists are suggesting that the theory of natural selection will help us to explain both the mind and social behaviour. The idea is not new: Darwin himself argued that there was no good reason to suppose that natural selection was responsible for all of life except human thinking. Following Darwin, the key idea of evolutionary psychology is that the mind is a collection of evolutionary adaptations, just like the body. Each 'mental adaptation' has by definition evolved because of its beneficial effect on individual fitness. Because the human environment is so greatly dominated by other humans, and thus by the difficult dynamics of group interactions, we can expect the psyche to have evolved into a finely tuned system for resolving the conflict between individual and societal organisation.

What does this mean in practice? Evolutionary psychology often looks like an attempt to apply biological theory to the discontents of society: stress, sexual harassment, homicide. One particular interest is the question of sex differences in humans. There is no doubt that men and women have physical differences that are determined biologically, but what about psychological differences? There has been a great deal of debate about the idea that men and women think differently. This is an area where evolutionary psychologists believe the theory of natural selection is decisive: that male and female psyches, as selected by the rigours of mate selection and parenthood (motherhood in particular) in ancestral society, have evolved differently.

The argument goes something like this. A woman, by virtue of carrying the eggs, and then the fetus, invests more in the offspring in the short term than does the man. The man's role, however, is important. For a woman (remember we are talking about a hypothetical ancestral society where the social services did not exist), it pays to find a mate who is healthy and socially powerful, and likely to stick around long enough to support the offspring. Put into the language of popularised evolutionary psychology, a woman will be choosy about who she goes to bed with: to sleep around is to risk being inseminated by a low-value man unable to support a child.

Men, however, have a different biology. Their parental investment starts with providing one sperm, out of the many million they manufacture daily. In principle, their genes are replicated most effectively by having as many sexual partners as possible. What might, therefore, be a biological reason for promiscuity is offset by another male necessity, one more precisely symmetrical with the lot of the female: the need to find a mate whose ability to nurture offspring, and not be tempted away, is demonstrably clear (by her history of maidenly virtue). Men's psyches have evolved an interest in promiscuity, and are riddled with fantasies about sexual relations without emotional ties; but they have also evolved psyches that are adapted to recognise the need for a reliable mate for the long-term nurture of their children.

These, crudely, are the predictions that evolutionary psychologists give us when they turn to sexual behaviour. The predictions spring from Darwin's theory of natural selection, and from the assumption that the psyche evolved in particular ways in ancestral society, to give adaptations that survive (however politically incorrect they may be) to this day.

Many surveys have been done to test the correctness of these predictions. Typically, men and women students are interviewed on campus, with anonymity preserved. Asked how many sexual partners one would ideally have in a year, male students favour eight, female students one or two. The 'discovery' of such a difference is described by evolutionary psychologists as an evolved sex difference. The psychologist David Buss, in an article in the social policy journal *Demos Quarterly*,[9] discussed these trials and explained: 'These findings ... confirm that men, on average, have a greater desire for sexual variety than do women – a straightforward prediction from the theory of parental investment and sexual selection'. The Darwinian scholar Helena Cronin, interviewed in *The Philosophers' Magazine*,[10] puts it this way 'Men can get away with the briefest of encounters; women are committed to nine months hard labour, nutrient-rich milk, unceasing vigilance.'

This argument relies in the first place on the idea that psychological sex differences between men and women are adaptive for each of the genders, and arise from the male and female reproductive strategies. No doubt this could be disputed, perhaps by showing that male and female reproductive strategies in nature are by no means as simply demarcated as the evolutionary psychologists imply. This would be an empirical matter, in other words something to be tested by observation. However, there is also a problem about definitions: is it correct to consider mating behaviour in birds of paradise as in the same category as sexual behaviour in humans? The latter, surely, is much more socially affected. For scientists there may be something attractive in reducing human behaviour to a few simple descriptions, such as 'men are naturally promiscuous', but simplicity is not a guarantee of truth. For instance, if in one interpretation Darwinian biology predicts male promiscuity; and male students declare under a cloak of anonymity that they would like to sleep with as many women as possible, and if also, amongst animals females are more active in looking after offspring, even together these details do not reliably tell us very much about how men and women behave, or should behave. You would probably find out more by reading a novel by Jane Austen.

The book *A Natural History of Rape* by the US academics Randy Thornhill and Craig T. Palmer is an extreme example of how evolutionary arguments are sometimes applied to human behaviour.[11] The authors write:

[9] D.M. Buss, 'Vital attraction', *Demos Quarterly*, **10**, 12–17.
[10] Issue 11 (Summer 2000).
[11] *A natural History of Rape: Biological Bases of Sexual Coercion* (Cambridge, Mass.: The MIT Press, 2000).

'Rape should be viewed as a natural, biological phenomenon that is a product of the human evolutionary heritage.' Within the field of evolutionary psychology there are differences of approach; Thornhill and Palmer's views may not be mainstream. On the whole, evolutionary psychologists are careful not to condone promiscuity, or its variant, sexual harassment, but they do urge that these behaviours are natural (while stressing that natural does not mean 'good'). There is a message for politicians and policy makers too. When legislating, or judging behaviour, or trying to improve it, do not ignore Darwinian processes. We might legislate against sexual harassment, and make it equally punishable, whether committed by a man or a woman, but according to the world view of evolutionary psychology, that would be to ignore evolutionary fact.

Research ethics

... as laboratory life has become more competitive, and especially where experiments are difficult to replicate, fraud and other types of serious misconduct have become less rare.

Nature, 4 March 1999, p. 13.[1]

8.1 A conflict of interest

An aim of this book is to show the way that biology is entangled in wider issues, such as politics or ethics. Some of my examples describe events occurring decades or centuries ago. Others are more contemporary: BSE for example, or the Human Genome Project. In these analyses, my emphasis is always on exploring the consequences of a simple fact: biology is a human activity. The controversies and the confusions that make biology so interesting do not result simply from technical inefficiencies. I have tried to repeat the point that the development of science is best understood as a consequence of the motivation and ambition of people and their institutions, as well as of their skills and their machines. Though I have emphasised the human side of science, and have mentioned by name many scientists, I have perhaps promoted the idea that individual scientists are somewhat passive, that they are reeds bending in various cultural, and professional winds. I now aim to explore the contrary idea: that scientists have a sense of responsibility, can stand up for what they believe, and need not follow the herd. The themes of our discussion will here be taken from the

---

[1] A. Abbott with R. Dalton and Asako Saegusa, 'Science comes to terms with the lessons of fraud', *Nature*, **398**, 13–17.

fast-growing field of research ethics, which considers how individual scientists should conduct themselves in their daily professional life. In this chapter, we get personal.

When the moral obligations of scientists are discussed, biology usually gets special mention. This is partly because, recently, the developments of biotechnology have raised questions about the implications of research. I have already covered some of these ethical aspects in Chapter 6, which discusses the Human Genome Project. However, the world of big business is another reason why research ethics have particular significance for the contemporary biologist. Biotechnology companies have attracted financial investment. Their research, perhaps using genetic technology to find new drugs, has obvious attractions for the stock markets of the world. The result is a close relationship between the culture of research and the culture of money. Commonly, announcements about the scientific research in a biotechnology company are released to the money markets before they make their way into a scientific journal; this happened, for example, when PPL Therapeutics cloned genetically modified pigs suitable for xenotransplantation. The practice represents a significant development in how science is communicated. It also raises the question: can a scientist in such a company ever get caught between the forces of finance, and the habits of scientific integrity?

We can imagine how such a conflict of interest could arise. Following the release to the press of promising results from clinical trials carried out by a private company, the market value of the shares of that company undergo a huge increase in value. The company's top clinical research scientist publicly questions the reliability of the company's rise in fortunes, on the grounds that the results of the trials were not, in his opinion, as promising as suggested in the press releases. The company executive stands firm with the opposite view, refusing to consider that there is any real evidence for the research scientist's doubts. Subsequently the research scientist loses his post. Whatever the merits of the two positions, we can see the problem. When a company's product is knowledge, some versions of that knowledge will be more saleable than others. In the case of a science company, there is a risk that the scientists know what kind of results will make good news, even before the trials or experiments have been performed. This does not mean that it is inevitable that companies, or their scientists, will act unethically, but it does mean that everyone needs to be aware of the possible existence of any conflict of interest, and be prepared to deal with it.

8.2 A code of conduct

It is common for professionals to have codes of conduct governing their behaviour. The UK press, for example, agrees to follow a voluntary code laid down by the Press Complaints Commission. In the case of science there exist firm regulations that bind scientists' actions: health and safety laws cannot be ignored; clinical research is scrutinised by ethics committees; legislation controls what kind of vivisection is and is not permitted. However, these rules are very specific, in contrast to the codes that discuss the ethical commitments of scientists and engineers.[2] Yet there is no doubt that the debate about ethical behaviour concerns in the scientific community is becoming more urgent. On 21 February 2002 *Nature* magazine published a commentary article doubting that the credit for scientific work is attributed fairly. The writer, Peter A. Lawrence from the MRC Laboratory of Molecular Biology at Cambridge wrote: 'Students are like boosters on space rockets, they accelerate their supervisors into a higher career orbit, and, when their fuel is spent, fall to the ground as burnt-out shells.'[3]

Though the problems may seem more urgent, the issues are not particularly new. For example the philosopher Karl Popper was clear that a code of conduct was necessary. However, writing in the 1970s, his worries were not so much about justice in laboratories, as about the wider social implications of science. For Popper, the question of the moral responsibility of scientists arose not because of sharp practice in claiming credit, but because of the fears of nuclear and biological warfare (this was the time of the Cold War). He saw that science had changed: 'all science, and indeed all learning, has tended to become potentially applicable. Formerly, the pure scientist had only one responsibility beyond those which everyone else has – that is, to search for truth'.[4] Popper himself put forward a scientists' version of the Hippocratic Oath, the ancient code that medical students would sign up to on becoming a doctor. He suggested that the primary moral responsibilities of the science student are: (1) to search diligently for the truth; (2) to respect his or her teachers, without slavishly following intellectual fashion; and (3) to show overriding loyalty to mankind.

[2] There are more codes written for engineers than there are for scientists. The Online Ethics Center for Engineering and Science at Case Western Reserve University (http://www.onlineethics.org/codes/index.html) lists a large number of such codes of conduct, most of them rather general in tone: 'Engineers shall associate only with reputable persons or organisations' is one recommendation of the American Society of Mechanical Engineers.
[3] 'Rank injustice: the misallocation of credit is endemic in science', *Nature*, **415**, 835–6.
[4] Karl Popper, *The Myth of the Framework: In Defence of Science and Rationality*, edited by M.A. Notturno (New York: Routledge, 1994).

These are big sentiments, and mean very little unless some detail is added. Yet it is interesting to see the conflict between Popper's code, and the atmosphere evoked by the idea of 'rocket booster' students. For if science stands accused of exploiting students, then it is not enough merely to put ethics courses into postgraduate training programmes.

In his book *The Ethics of Science*,[5] David B. Resnick attempts to describe a code of conduct that would deal with the concerns raised by Karl Popper, as well as with the contemporary pressures currently bearing down on science. Here are some of his recommendations:

1 Scientists should not fabricate, falsify or misrepresent data or results.
2 Scientists should avoid error, self-deception, bias and conflicts of interest.
3 Scientists should share data, results, ideas, techniques and tools.
4 Scientists should be free to conduct research on any problem or hypothesis. They should be allowed to pursue new ideas and criticise old ones.
5 Credit should be given where credit is due but not where it is not due.
6 Scientists should educate prospective scientists, and inform the public about science.
7 Scientists should avoid causing harms to society and they should attempt to produce social benefits.
8 Scientists should not be unfairly denied the opportunity to use scientific resources, or advance in the scientific profession.
9 Scientists should treat colleagues with respect.
10 Scientists should not violate rights or dignity when using human subjects in experiments.

This detail of the code, and its implicit acknowledgement that contemporary science is aggressively competitive, signal that it is less certain than before that ethical behaviour can be taken for granted. You will see that the list is not in order of importance: few scientific wrongs could be worse than the violation of human rights during experiments (point 10); and traditionally, scientific wrongdoing has been conceived as largely consisting of fabrication, falsification and plagiarism (point 1). Overall, the list gives one an uneasy sense that science is now contaminated by malign influences of commerce, personal insecurity and ambition. Yet was there ever a golden age, and if things have slipped since that time, how bad has it got? When *Nature* magazine published Peter Lawrence's indictment, it asked

---

[5] London: Routledge, 1998.

various of its advisors for their opinions.[6] One wrote: 'There are bad apples in the scientific community to be sure. And there are poor mentors, and serial abusers of younger scientists for the advancement of the 'Great Leader'.... Yet the majority of scientists and mentors are not like that – so my fear is that the majority could be tarred by the sins of the minority.' This comment, with its mention of 'bad apples' and 'sins' suggests that there is a hard core of 'scientist-criminals' who contaminate the body of science. Yet the evidence is that another kind of analysis would be more appropriate. Rather than concentrating on rooting out and expelling a few fallen souls, the emphasis might be better placed on understanding how the pressures of contemporary science might force almost anyone to cut corners, or artificially boost their standing. This seems to be the hidden message of Resnick's code. An editorial in *Nature* magazine in 1999 put it this way: 'there is also need for a systematic study, by surveys if necessary, of issues of credit misappropriation, exploitation and inappropriate pressures arising in laboratory hierarchies'.[7] The quote at the head of this chapter makes a similar point.

To identify bad behaviour, you first need to know what you would expect to see if all is going well. Commonly, the resource used at this point is a paper by the Amercican sociologist of science Robert K. Merton: '*The Normative Structure of Science*'. This was first published in 1942, and is famous for its attempt to lay down the ethos – the values and the norms – of scientific research.[8] As you can imagine, the age of this work immediately raises questions over how well its ideas transfer to a new millennium. In fact, 'Merton's Norms' still get good coverage. The reason may be that he paints a picture of science that is more than positive – it is inspiring. In the space of ten elegantly written pages, Merton catalogues the good habits of scientists, and implies that only the actions of an autocratic society could seriously threaten the harmonious balance of the scientific culture. There are four norms in Merton's framework. The first is universalism, according to which the findings of science depend not at all on the race, nationality, class, or particular qualities of the individual scientist. All that matters is the quality of the work. As Merton puts it: 'an anglophobe cannot refute the law of gravitation'. The second norm is communism. Science is done

[6] 'Thoughts on (dis)credits', *Nature*, **415** (21 February 2000), 819.
[7] 'Where next on misconduct?' *Nature*, **398** (4 March 1999), 1.
[8] 'Science and technology in a democratic order.' *Journal of Legal and Political Sociology*, **1** (1942), 115–26. Republished as 'The normative structure of science' in *The Sociology of Science: Theoretical and Empirical Investigations*, edited by Norman W. Storer (Chicago: Chicago University Press, 1973), 267–78.

by people within a community, working in collaboration, and with the spirit of openness and sharing. The results of that work belong not to an individual, but are 'assigned to the community'. Third, scientists are distinctive for their disinterestedness (the term refers not to a lack of interest, but a lack of bias or partiality). For Merton, the activities of scientists are 'subject to rigorous policing, to a degree perhaps unparalleled in any other field of activity'. The specific point is that scientists do not feel any particular attachment to their theories or results, certainly not to the extent that they would defend their work more than is reasonable. Conversely, nor would they attack the work of others, motivated simply by the fact that they might be competitors within the same field. Finally, Merton suggests that the fourth norm is organised scepticism. Here he refers to the way scientists have procedures for vetting suggestions in a fair and orderly manner, subjecting theories to a courteous scrutiny that will ensure that poor thinking simply vaporises. Merton is referring to the constant discussion of scientific conferences, to the journals that form the written record of scientific achievement, and the various mechanisms (such as peer review) that ensure quality control of the papers submitted for publication.

As you can see, Merton's norms do not seem to fit well with the atmosphere evoked by David Resnick's code of conduct, or even with the various quotes from *Nature* magazine that I included earlier in this chapter. However, no one is suggesting that science has suddenly transformed into something not much different from share dealing, though, as we shall see, there have been striking cases of fraud in science, where the word used more commonly is 'misconduct'. This is is not simply a euphemism; it is an attempt to convey the idea that the moral standing of science is threatened more by the steady drip of minor 'misdemeanours' than by the sudden shock of major fraud.

The simple fabrication of results is certainly a rare event in science, and is considered a serious offence against the integrity of the scientific community. However, the distortion of results so as to fit in with a preconceived idea (whether of the scientist or the sponsoring company) is likely to be much more common. Data always have to be manipulated in order to reveal their meaning; the skill lies in minimising self-delusion, which occurs when your commitment to a project prevents you from giving due consideration to negative results and anomalies. At the end of an experiment you might simply decide that such results should be deleted – they are a result of 'experimental error', and of no interest to anyone. It would clearly be a more serious act to remove the data because you believe they undermine your theory, and perhaps your academic standing in the field. If you are the confident sort, with a few well-received papers in circulation, the upsetting results might be no more than an irritation. If however you feel unlucky and unappreciated, and are reaching the end of a 3-year contract, you might persuade yourself that the deletion is necessary and even justifiable, a case of 'me-against-them'; nor need this be an entirely conscious decision.

The problem of self-delusion arises simply because science, like art, is an interpretative activity. It is interested not simply in getting results, but in determining their meaning. This is why science is creative and technical. Reading too much into results, or simply ignoring inconvenient data, is a professional hazard that could, in some circumstances, lead to deception; those circumstances are most likely to arise when the environment is highly aggressive and competitive.

8.3 Peer review

I have argued, then, that to understand the significance of misconduct in science, you need to understand aspects of the institution of science, as well as of the psychology of scientists. Here, peer review, the accepted method of evaluating work in science, is particularly relevant. I will consider the issue only in relation to journals, because this is probably where most of the controversy lies. The procedure begins when a scientist submits a manuscript to a journal. The editor sends out copies to colleagues in the field ('peers') who have the expertise and, one hopes, the fair-mindedness to judge the work properly. The referees are asked to give an opinion, perhaps ranking the work on a numerical scale of merit, and recommending whether it should be published, rejected, or accepted providing certain modifications are made. When the referees' reports are received back in the journal office, the editor makes a decision, and informs the author (or authors), usually sending them the referees' views. Throughout this process, the authors are not given the identity of the referees (the principle of anonymity). The value of peer review lies in getting an outside, but informed, view on a piece of work; it is science's main act of quality control. However, there are also those who view the process with some cynicism, as this quote shows: 'There seems to be no study too fragmented, no hypothesis too trivial, no literature citation too biased or too egotistical, no design too warped, no methodology too bungled, no presentation of results too inaccurate, too obscure, and too contradictory, no analysis too self-serving, no argument too circular, no conclusions too tri-fling or too unjustified, and no grammar and syntax too offensive for a paper to end up in print.'[9]

Those who question the merits of peer review put forward a number of points. They question the impartiality of referees, who can use their anonymous reports to cast doubt on, or delay the publication of, the work of competitors. Critics ponder too whether papers that are unusual or controversial, and challenge some accepted wisdom, can receive fair treatment.[10] Conversely, those authors who are well known to the judges, and whose work is already respected, are alleged to receive favourable treatment. To the critics, the concern over peer review is not simply that it is conservative: far worse, it is an 'old boys network'. Clearly, for peer review to be effective, it must avoid not only favouritism, but sexism and nepotism.[11]

Some problems of peer review cannot be blamed either on editors or referees. For example research papers are sometimes 'salami-sliced'. This is when one big report, with plenty of results and proposals, is fragmented into several papers, and sent to a variety of journals. This seems to make good sense from the point of view of the author, who gains multiple publications, instead of just the one. However, the suite of publications is bound to include some redundancy and repetition. If the message of each publication is that a particular medical treatment is effective, a literature search will indicate greater professional backing for that treatment than if the results, and recommendations, had been confined to one journal only. Since doctors and health professionals use journals to guide their practice, such 'salami-slicing' (or 'disaggregation') could be construed as unethical.

Journal editors also worry about the question of authorship. It is common for a scientific paper to list a dozen or more authors. The question raised is: who did the work? Honorary authorship is the practice of listing as an author someone who did none of the work described, perhaps the team leader, or someone else who might one day return the favour.

[9] D. Rennie, *Journal of the American Medical Association*, 261:17 (5 May 1989), 2543–5; quoted in *Scientific Deception: an Overview and Guide to the Literature on Misconduct and Fraud in Scientific Research*, by L. Grayson (London: British Library (1995), p. 37.
[10] In one investigation, authors sent out two manuscripts for review, both of them fictional, both of them submitted for publication in a non-existent journal. One concerned a novel, but conventional medical treatment, the other a complimentary therapy. The latter paper received significantly worse reviews. See K.L. Resch, E. Ernst and J. Garrow, 'A randomized controlled study of reviewer bias against an unconventional therapy', *Journal of the Royal Society of Medicine*, **93** (2000), 164–7.
[11] See the commentary article by Christine Wennerås and Agnes Wold 'Nepotism and sexism in peer-review', *Nature*, **387** (22 May 1997), 341–3.

The favour consists of getting another paper on your CV.[12] The importance of these debates for research ethics is that peer review remains the main system for preventing scientific deception, whether the deception is in its mild, delusory, form or is a deliberate act. Once a paper is published, it is 'in the literature', and has leapt into the public domain. It will be read, referred to, used in teaching; and its authors will have their careers advanced by one notch. Journal editors can and do publish retractions, for example when a published paper is found to have included faulty data or incorrect experimental protocols, but by publishing such amendments, even when they come from the author, the journal inevitably sows suspicions in readers' minds, suspicions that may quite unjustifiably blight the career of the author.[13]

The establishment's view of peer review is that, generally speaking, it works well, moreover, there is no better system available. Now that research ethics is a topical debate, journal editors often express views on the system, and sometimes suggest changes, for example that the referees should not do their work anonymously, but certainly no radical changes to science's pre-eminent system of quality control are imminent. It is important then that the actual environment in which scientists work, whether in universities or industry, is supportive of debate, and allows the exchange of ideas and the trusted reading of the rough drafts of colleagues' papers.

8.4 A historical overview

The debate over peer review reveals clearly one aspect of contemporary research ethics: problems over misconduct are a matter for institutional self-reflection, as well as for individuals. In order to emphasise the significance

---

[12] Citation statistics are compiled by the Institute for Scientific Information (ISI), a company in Philadelphia, USA, and are widely used as a measure of the quality of scientists' work. Doubts about the accuracy of this information are raised in *Nature*, **415** (10 January 2002), 101. *Nature* magazine has also attempted to lay down some policy on the matter of authorship. In their advice to scientists wanting to submit a manuscript, the editors write: 'Authors are encouraged to specify the contribution made by their co-authors in the Acknowledgements ... we hope that, as the practice spreads, the dishonourable practice of 'honorary authorship' – authorship by virtue only of seniority, for example – will diminish'.

[13] R. Horton, '*Revising the research record*', *Lancet*, **346**:8990 (16 Dec 1995), 1610–11. For a full listing of various publishing misdemeanours and ambiguities see the annual *COPE Report* (London: BMJ Books), which is put together as a co-operative enterprise by more than 50 editors of journals (the Committee on Publication Ethics, or COPE), each wishing to air some of their concerns about some of the manuscripts that cross their desks. The *COPE Report* is an excellent place to learn more of the successes and failings of the peer review procedure. See also the COPE web site at http//:www.publicationethics.org.uk, and the Online Ethics Center for Engineering and Science at Case Western University web site at http//:www.onlineethics.org.

of this, I will now describe the historical context to research ethics, and show how much ideas have changed in the last 50 years.[14] We can see three themes running through this history, each with a different emphasis of concern, each responding to novel interactions between science and society: these are at the level of the individual, society, and science itself.

Theme one: saving the individual

We saw in the last chapter that eugenics raised many issues relevant to research ethics. The most obvious one refers to the rights of the individual. Eugenic laws that made sterilisation compulsory simply rejected the idea that a disabled individual has the right to reproduce. As we have seen, under the Nazis this branch of biology mutated into a campaign of murder. Scientists, doctors, nurses, engineers and architects were just some of the professionals who had broken the fundamental rule that science and medicine is aimed at improving the human condition. Therefore, after World War II, it was clear that science in general, and biology in particular, needed to face up to Nazi atrocities. A series of trials after the war, held in Nuremberg, Germany, put some of the war criminals on trial, and gave rise to an early statute in research ethics, called 'The Nuremberg Code'. This examined the issue of experimentation on humans, and tried to establish norms that would prevent science deliberately hurting humans, even if some 'greater good' could be argued (such as the development of a new drug). Nazi science showed that scientists could not in all circumstances be depended upon to act on behalf of the individual. Therefore, the code states that no one should be coerced to take part in an experiment, and that once enrolled on a trial, the human subject must have the right to withdraw at any time. Clearly, for such participation to be truly voluntary, the subject must not feel under pressure, must be briefed about possible harmful effects, and capable of understanding the scientists' intentions.

The revelations about Nazi science had no particular effect on general attitudes to science and scientists. Customarily, we see the period 1945–66 as a time of public benevolence towards, and excitement in, the achievements of science. Science and technology formed a key element in the

[14] This section on the development of research ethics draws on the ideas of Stanley Joel Reiser ('The ethics movement in the biological sciences: a new voyage of discovery' in *The Ethical Dimensions of the Biological Sciences*, edited by R.E. Bulger, E. Heitman and S.J. Reiser (Cambridge: Cambridge University Press, 1993). A second edition of this book was published by CUP in 2002, under the title *The Ethical Dimensions of the Biological and Health Sciences: A Textbook with Case Studies*.

rebuilding of Europe. In the main, it seems, scientists were trusted: no doubt an important aspect of this was due to the Nuremberg Code, which had exorcised the Nazi demons from the practice of science. It was during the 1960s that the steadiness of public trust began to weaken. In 1963 there was great interest in a case where three physicians, as part of an experiment, injected live cancer cells into 22 sick and elderly patients at the Jewish Chronic Hospital in Brooklyn. There was no consent. When this was later challenged in court, the physicians' work was upheld because it was deemed an experiment, rather than a therapeutic encounter. In other words, because the work was considered scientific rather than medical, the existing ethical codes did not apply. Not surprisingly, controversy ensued, and guidelines were drawn up to ensure that the consent of subjects would always be sought in future. Institutions themselves had to set up ethics panels, made up of scientific peers and, significantly, lay members of the community.

The literary periodical *Granta* has revisited these issues of consent in human experimentation by examining a set of famous psychology experiments that investigated the tendency of people to conform.[15] Working in the early 1960s, Stanley Milgram gave his subjects instructions to interrogate another person (actually an actor) behind a screen. The instructions included the demand that the subject give electric shocks whenever a wrong answer was made. The actor gave incorrect responses, was given simulated electric shocks, and groaned, cursed, screamed and asked for mercy; the sounds were piped back into the room where the subject sat tensely over his equipment. Whenever the subject showed anxiety at the fate of the victim, a white-coated technician intoned 'The experiment requires that you carry on'. The experimental results were interpreted as suggesting that the average American is basically a torturer. Milgram was surprised by how long it took before the subject withdrew. Yet very quickly the work was denounced as unethical, and against the spirit of the Nuremberg Code. For though the subjects were debriefed after the experiment, a number of them found the experience traumatic and destabilising; one returned to Milgram's office, complaining of heart problems. Ironically, Milgram had ensured the fame of his experiment by describing it as relevant to the issue of Nazi atrocities, indeed as in some way 'explaining them'. His career foundered, and he died surprisingly young. Yet, whatever the doubts of the academic community, psychology dons

---

[15] I. Parker, 'Obedience', *Granta*, No. 71 (Autumn 2000), 99–125.

for many years made good use of the Milgram experiment (his conclusions, not his downfall) as an eyes-wide, shock-horror introduction to undergraduate courses in social psychology.

Theme two: saving society
Another aspect of research ethics concerns not just the safety of individual, but the safety of society itself. Once again, World War II was an important prompt for these worries. During the war, many scientists were put to work on new technologies ranging from explosives and munitions to radar and encryption. For most American and British scientists, this work was ethically uncomplicated. There was an overriding moral necessity to defeat the Nazis: it was part of the war effort. However, one device developed during the War came to be symbolic of the moral dimensions of modern science: the atom bomb. In the USA, this started in late 1941 as a secret programme, which came to be known as the Manhattan Project. It is reported that the physicists working on building the first atom bomb (a fission bomb) in Los Alamos, enjoyed their work. They were in the company of the world's best physicists. They were engaged together on a single project at the cutting edge of particle physics, and faced with the possibility that Hitler himself might be building such a bomb, the moral arguments to do the work must have seemed compelling. A key moment in the story comes when the bomb was first tested, in the Nevada Desert, the so-called Trinity Test. The scientists travelled by jeep to the test site, set up camp 10 kilometres from the blast tower, and put on sunglasses. No one was quite sure whether it would work. The night before a couple of scientists had wondered whether the force of the blast might ignite the atmosphere; there was general uncertainty about how strong the explosion would be. When the bomb detonated, according to plan, at five o'clock in the morning, most of the watching scientists reportedly expressed quite bland sentiments, such as 'Well, it worked'. More famously however, the director of the Project at Los Alamos, Robert Oppenheimer, quoting the Hindu god Vishnu said 'Now I am become Death, the destroyer of worlds', while Ken Bainbridge, the test director, told Oppenheimer 'Now we're all sons of bitches'.

During the 1950s and 1960s the Cold War led to the build-up of stockpiles of nuclear warheads in both the Soviet Union and the NATO countries. The destructive capacity of these weapons was such that no war between the Soviet Union and the USA could have a victor. The acronym of the time, MAD (Mutually Assured Destruction) summed up how strategically inflexible weaponry had become. Politicians alleged that the weapons could never be used, but were nevertheless necessary for purposes of deterrence. The insanity of the situation eventually provoked protests on university campuses and in cities, organised for example by the Campaign for Nuclear Disarmament (CND). CND included a number of scientists; and there were many scientists, including physicists, who made plain their opposition to the arms race, though their powers to do anything were negligible. Bomb production was in the hands of the military; arms limitation talks were organised by the politicians. The time when nuclear scientists could influence the use of their technology was long gone. There is a story that when Robert Oppenheimer said to President Truman that 'I feel we have blood on our hands', Truman replied 'Never mind, it'll all come out in the wash', and instructed his aides to make sure that Oppenheimer did not come visiting again.

In the 1960s the biologists, less visible than the physicists in the three decades after the war, acquired new powers with the rise of molecular biology, and especially of genetic engineering. Now it was their turn to consider the social implications of their work. In particular, a few biologists wondered whether experiments involving the insertion of foreign genes into *Escherichia coli* should be sanctioned. The concern was that *E. coli* could grow in the human intestine, and that the viruses used to transport genes into *E. coli* were capable of causing tumours. With these technologies now becoming available, some biologists felt as though their science had acquired powers analogous to those of the scientists of the Manhattan Project – and that scientists should pause for a while to consider the social implications. Unlike the situation in the Nuremberg deliberations, where it was the rights of the individual that were considered, the focus was now on the mass population, at risk of annihilation by an outbreak of a genetically engineered pathogen: it was a safety issue, not a question of whether it is right to genetically engineer an organism. As a result of these concerns, a conference in 1974, at Asilomar in California, reviewed these new biohazards, and drew up a set of guidelines for ensuring the safety of genetics experiments. In one sense, these discussions simply reflect the proper concerns of biologists for the safety of their colleagues and the mass population; but just as importantly, the debate showed biologists taking responsibility for the wider implications of their work. They looked beyond the small arena of the Petri dish and the published paper, and saw that a consideration of the benefits and harms of biological science was part of their job too.

Theme three: saving science

Another phase in the development of research ethics arose in the 1980s and 1990s. This was the era of greater involvement by industry in funding science, as science has become more expensive. Many biologists have a stake in biotechnology companies, and science is now thoroughly entangled in the private sector. At the same time science has received more intense scrutiny by governments, which have also been spending more on research. In 1981 a US Congress committee chaired by Congressman Al Gore looked at the question of fraud in science. Government interest had been provoked because of one or two cases of misconduct: could the scientists be trusted with government money? Gore wondered whether the few cases his committee considered might be just the tip of the iceberg, and asked whether scientists were capable of dealing with the ethical judgements that now confronted biology. Not surprisingly, given the heightened awareness produced by the Gore investigations, several cases of alleged misconduct in the 1980s, involving the fabrication of data, drew the attention of the press. In 1981 a Harvard cardiologist John Darsee had admitted falsifying scientific data. After Darsee had been disciplined, attention shifted to the 47 co-authors listed in his 109 papers. An investigation by two scientists, Walter Stewart and Ned Feder, free-lancing as fraud busters, suggested that some of the co-authors had failed to spot errors in Darsee's papers – the implication being that they had not been involved in the work, and had not even read the manuscript. This evidence of the nature of 'honorary authorship', tied to out-and-out fraud, was a severe embarrassment. If the published paper could not be trusted, what implications did this have for the process, and the reliability of science?

Then in 1985 another case became public, and remained argued over for at least a decade. The so-called Baltimore Case had at its centre the Nobel Prize winner David Baltimore, immunologist and president of the Rockefeller University. His name appeared on a paper produced by a member of his team, Theresa Imanishi-Kari. According to a junior scientist working in the laboratory, Margaret O'Toole, the data in this paper had been fabricated, and she 'blew the whistle', alerting several senior colleagues. In due course the case found its way to a government committee, and because Baltimore was a Nobel laureate the case made headline news. The US Congress became involved; laboratory papers were subpoenaed by the secret service; notebooks were X-rayed. In the end Baltimore was cleared of all wrongdoing, but in the meantime most people were surprised to learn that the named authors of scientific papers often had only had the most oblique relationship to the actual work – perhaps as little as an encouraging conversation in a lift.[16]

8.5 Theories of ethics

No one knows how common misconduct might be. In this chapter I have quoted frequently from *Nature* magazine because that seems a good way of indicating that many scientists consider the issues important ones. However, identifying a problem is one thing, solving it another. The problems of research ethics seem particularly complex because, if my interpretation is correct, we are dealing here with issues that are as much institutional as they are personal. In other words, though no doubt there are 'bad people' in science, just as in any profession, simply targeting them is not a complete answer.[17] As I showed with the debate about peer review, there are institutional matters to be looked at as well.

Internal reform, carried out rigorously, could minimise the following examples of sharp practice: sexism and nepotism in peer review; honorary authorship; exploitation of students; undisclosed conflict of interest; 'salami-slicing'; and the secrets and lies sometimes associated with some commercial priorities. It would be a policy partly of bureaucratic rectitude, partly of prohibition. No doubt the policy would be worthwhile, but it would have a legalistic, prohibitive feel; it would be associated with the proposals of funding bodies, journal editors and institute directors. Moreover, policies that aimed to rid science of misconduct by reform of its institutions need not be described as 'ethical', but as simply a matter of justice, or even of efficient organisation. Perhaps, then, it is a misnomer to talk of research ethics. If you consider the themes of this chapter, they have been more concerned with the way an institution might encourage misconduct, than with what might be meant by 'good' or 'bad' character. Yet throughout this book I have been arguing that the well-trained biologist must be someone who is able to reflect on the historical, philosophical and political contexts of the subject. I have been urging that biological

---

[16] For stories about misconduct, see David J. Kevles' detailed *The Baltimore Case: A Trial of Politics, Science and Character* (New York: W.W Norton, 1998). *The Undergrowth of Science: Delusion, Self-Deception and Human Frailty* by Walter Gratzer (Oxford: Oxford University Press, 2000) is a more popular account of stories of misconduct.

[17] Is science a place where dishonest people could flourish? While science is very often a team activity, making individual deceit and secrecy rather obvious, for short term gain, simple fabrication of results would be relatively easy – for the period of time you got away with it, the abuse would be one of trust. Fabrication of results is a serious blow against science because others, who unwittingly use your results, are drawn into the intrigue, and ultimately have their work invalidated too.

training should include experience of standing back from the edge of research, to look instead at perspectives.

The people who founded ethics, the Greeks, argued that the study of ethics should be the study of peoples' characters. We saw this right at the start of this chapter, where I described Karl Popper's suggestion that scientists take a version of the Hippocratic Oath. There we saw that loyalty to your teachers, and to the search for knowledge, are paramount. It is personal qualities that are stressed, not institutional ones. Yet the question of personal quality – of being a person of excellent character – is far from the focus of any scientific training. The state of your character, and the state of your scientific prowess, are in modern society entirely separate issues. It is not difficult to conceive of someone being a good scientist, but a bad person, and as we have seen in this chapter, it is easy to have a discussion of research ethics without reflecting on the kind of personal qualities that we should expect from our scientists. We need, therefore, to take a brief look at the kind of ethics theories in current use, and to evaluate them for the contribution they can make to evaluating rights and wrongs in science, and for the guidance they can give to individuals as well as institutions. I will conclude this chapter by examining three theories of ethics. The three theories are duty-based ethics (sometimes called deontology), utilitarian ethics (sometimes called consequentialism) and virtue ethics. I will suggest that it is virtue ethics, which explicitly emphasises the notion of excellence of character, that might well provide us with new ideas on what we mean by a 'good scientist'.

Deontology is the kind of ethics where it is rules and regulations that dictate your moral decisions. There is a list of duties that you simply must follow: to transgress them is to act immorally. Furthermore, such duty-based ethics are based on the idea that you must never break the rules, even if the consequences of following the law are, in a particular instance, very regrettable. Thus a deontologist who followed the rule that you should not lie would tell the truth even if such honesty got a friend into trouble. Christian ethics are a good example of a rule-based moral programme, especially if you take a literal view of the Bible, and for instance focus on the ten commandments. For a deontologist following Christian teaching, the commandment 'Thou shalt not kill' is an absolute decree, not to be broken under any circumstances (even if the proposed murder victim is a child-eating psychopathic dictator). We can see several problems with all this. The first is that deontology relies on an authority – in the case of Christianity, holy scripture – that not everyone recognises. The second is that rules tend to ignore the subtle complexities of real life; Christian theologians, for example, spend time debating when a war might be just. Third, deontological ethics seems to suggest that one can be a morally good kind of person, merely by obeying rules to the letter. There seems no room for personal decision. What is right or wrong is decided for you in advance, simply as a result of the religion, profession or other group you belong to. Though some people might find this attractive, in that moral responsibility has been taken away from the individual, and located in some absolute doctrine, others will find it worrying that people could ever want effectively to shut themselves out from ethical debate, and become a mere rule follower.

The debates over research ethics sometimes sound highly deontological. If there was a code of conduct for scientists, it would presumably be full of prohibitions such as 'Thou shalt not photocopy and distribute manuscripts sent to you for refereeing'. It would be a negative, legalistic code, one that lays down what should not happen, and implies that you become an ethical scientist by following this external code, not through any special initiative of your own. We can see deontology in the Nuremberg Code, which lays down rules on human experimentation; and in the laws on animal experimentation, controlling such matters as the use of anaesthesia. Again, the problems with deontology seem only too obvious. The rules about animal experimentation may be clear, and designed to eliminate 'unnecessary' suffering; a scientist might argue that, because he has followed the rules, he has by definition behaved ethically; but an antivivisectionist would not accept the validity of these rules, and so would consider them irrelevant in judging the ethics of a scientist.

Utilitarianism is by far the most common type of ethics encountered in science. This is the philosophy, advocated especially by John Stuart Mill (1806–73), that suggests that the moral value of an act is to be determined by its consequences for our pleasure or happiness. Thus, you might hurt someone now, and be morally in the right if, by consequence, the sum total of happiness in the world is increased. As we saw in Chapter 4, Section 4.4 (Animals have rights), utilitarianism is the routine justification for animal experimentation, because it is argued that the sum total of human happiness produced by new treatments, or safer cosmetics, outweighs the loss of happiness endured by the animals themselves. The same argument is used to justify taking money from the taxpayer, and funelling it into the science funding agencies such at the Medical Research Council (MRC) or the Biotechnology and Biological Science Research Council (BBSRC).

Utilitarianism has been defended as an ethical programme by point-ing to its lack of reliance on external rules, or on God. It has a simplicity that could be appealing. Utilitarianism, unlike deontology, offers a person some initiative in considering how to act. It is not rules that are important, but the consequences of what you do. However this simplicity is of course it downfall. By suggesting that such a thing as 'happiness' or 'pleasure' could be a common currency – that it can be easily measured, compared, and weighed up – utilitarianism puts itself in a tight corner. Consider the case of a war conducted by Western powers against a distant and repres-sive regime. In such cases we are used to hearing about 'collateral damage', that is, the death of civilians caught up in the bombing or the cross fire. We are also used to the utilitarian justification: such deaths, though regret-table, are acceptable because in the long run the ruling despot is destroyed, and global security is increased. How well does this argument work if it happens that amongst the casualties lie your parents?

If it is the change in total happiness or pleasure that is the most im-portant way to judge the moral worth of an action, then presumably this would be an argument for any strategy that made gluttony, drinking and sex a more central part of our lives. Yet these appetites need not not be plea-surable to everyone: some would regard them with horror. J.S. Mill him-self worried that utilitarianism could be used to justify happy and well-fed slothfulness, and suggested that the pleasures he considered worthy of measurement would be the 'higher pleasures'. He meant the kind of pleasures induced by a good book, or a fine concert. It hardly needs to be pointed out that this too is a matter of opinion, and simply sounds elit-ist. Finally, in addition to defining happiness, and measuring it, there is the problem of how far in the future you should go in measuring the con-sequences of an action. Perhaps a cure that emerged from animal experi-mentation in 5 years, is in 10 years superseded by another cure that was developed and tested without the use of a single animal.

You can see that deontology and utilitarianism are each important in establishing the ethical framework of science, but nonetheless have im-portant problems. A particular characteristic is that both downplay the role of the individual. Deontology simply requires the individual to obey rules, while utilitarianism only demands that the individual can judge consequences. So that leaves virtue ethics, the philosophy that originated with Socrates, Plato and Aristotle, which examines what might be meant by a person of excellent character, or of 'great souledness'. In an attempt to be more precise about the nature of this excellence of character, Plato described four central virtues. They were courage, which is the ability to control your fear; wisdom, which is the self-knowledge that allows you to make good decisions; temperance, which is the ability to control your desires; and justice, which is the awareness of the sensibilities and needs of others. The idea is that ethics should be centred not on acts or on rules, but on each individual's attempts to embody the virtues. In these attempts, the way you see your role in society – your profession – acquires special importance. Your professional self, and your personal self, become one. Thus, a soldier will acquire the virtue of courage as a result of his training; a prince should be practised in delivering justice before he becomes king. For the Greeks, this linking of the virtues to a person's role leads to the impossibility of saying something like: 'He was a good man, but a poor (cowardly) soldier.'

This linkage between personal excellence of character, and professional ability, is not one that we necessarily embrace in contemporary culture. The great scientist may not, in truth, be a person of 'great souledness', but it is interesting to speculate on how one would draw up a list of the virtues of the good scientist. We would not be interested simply in scientific skills, or background knowledge. Nor would ambition, nor competitiveness, be enough on their own. For the list to resonate with the virtues described by the Greeks, we would have to find characteristics such as self-reflection (which produces wisdom), honesty and trustworthiness. We saw earlier that when Robert Merton described his norms of science, such traits figured strongly; but we have also seen that as the twentieth century ended, success in science became more obviously associated with competitiveness and even secrecy. A debate is needed amongst biologists on these issues, one that will reflect as much on education as on the way in which laboratories approach their work.

Index

Page numbers followed by 'n' refer to footnotes